W9-AQN-669

Praise for

BACK TO EARTH

"Stott conveys the importance of space exploration with an astronaut's understanding and an artist's eye. Her unique perspective can help all of us better appreciate our home as well as encourage the passion needed to take care of it."

—Mike Massimino, astronaut and *New York Times*–bestselling author of *Spaceman*

"A smart, witty guide to living on our home planet from one of the only humans to have lived in space and under the sea, Stott understands our worlds, from algae to stardust. Now she lets us in on the magic. Buy, read, act!"

—Lynn Sherr, ABC News correspondent and author of *Sally Ride*

"I shared a delicious curry meal off-planet with Stott on the International Space Station and it felt just like home. When we landed and broke bread again I had a new perspective and appreciation of what it meant to be home. Through *Back to Earth*, Stott helps us all find that perspective and connection to our home planet."

—Leland Melvin, astronaut and author of *Chasing Space*

"Stott's wonderful book maps her extraordinary journey from Earth to space and back again. She teaches us how to see our world as the spaceship it really is, and how important it is to take great care of the round blue vehicle in which we all ride."

—Peter Gabriel, musician and humanitarian

“*Back to Earth* should be required reading for anyone who asks: Why do humans go to space? I have not seen a better description of the balance between the current risks and the beauty of life on our planet. Stott has succeeded in bringing the astronaut experience to the heart of the reader, sharing not only her stories but also her optimism for a promising future for life on Earth.”

—Eileen Collins, colonel, USAF (ret.), astronaut, space shuttle commander, and pilot

“A thought-provoking and thrilling call to action to do right by our environment and, even more so, by each other.”

—Scott Harrison, *New York Times*–bestselling author of *Thirst*

“Nicole inspires us to be a crew member, not just a passenger, on the most important spaceship of all: planet Earth.”

—Simone Giertz, inventor and YouTuber

“Through the windows of the space station, floating alongside our crewmates from around the world, Nicole Stott and I saw Earthrise together, and understood how fragile our water-filled home is. By turning her journey into a beautifully written work of art, Stott has taken action in a way that few can. Her book is very much needed.”

—Guy Laliberté, seventh space tourist and founder of Cirque du Soleil, Lune Rouge and ONE DROP

“Our planet is a spaceship, and we are its crew! This is the story of a down-to-earth astronaut who cares deeply about our planet and all life that shares it as home. Sometimes you have to leave a place to truly appreciate it—nobody tells this tale better than Nicole Stott.”

—William W. Li, MD, *New York Times*–bestselling author of *Eat to Beat Disease*

"In *Back to Earth*, Nicole Stott offers us the precious gift of perspective on human life. With wisdom gained through her unique experiences, Stott connects the dots between the heavens and Earth, creating an elegant template for living rich lives in harmony with nature. *Back to Earth* is the answer to 'They should have sent a poet.'"

—Brian Skerry, *National Geographic* wildlife photographer and film producer

"Sometimes you have to leave the planet to love the planet—and certainly to appreciate it fully. That, and much more, is the core message of this wise and artful book. What her trips to space taught her about survival, problem-solving, and our own humanity, Stott in turn teaches us—making us not just passengers on our planet, but crewmates. I came away from this book a better Earthling than I was when I began it."

—Jeffrey Kluger, *TIME* editor-at-large, coauthor of the national bestseller *Apollo 13*, and author of *Holdout*

"*Back to Earth* should be in everybody's home library! Stott has captured the incredible beauty, fragility, and interconnectivity of Earth, writing with the inspirational flair of Carl Sagan, rallying the reader to fully embrace their role as a guardian of the future. A very human story.... This is a book for the crew of Spaceship Earth."

—Mike Mullane, astronaut and author of *Riding Rockets*

"*Back to Earth* offers an extraordinary glimpse of Earth from a view most of us will never witness firsthand. Nicole Stott underscores the urgent need for humanity to take action, while offering hope that we can solve our greatest problems if we work together for the greater good."

—Leilani Münter, environmental activist, filmmaker, and retired race car driver

"Stott came back to Earth with a new mission: urging all of us to protect our one and only 'home.' As an artist, she conveys her spaceflight experience in her paintings; and in her wonderful book, she finds the words to share her profound and moving journey as a call to action for us all."

—Frank White, author of *The Overview Effect*

BACK TO
EARTH

BACK TO EARTH

WHAT LIFE IN SPACE TAUGHT ME ABOUT OUR HOME PLANET—AND OUR MISSION TO PROTECT IT

NICOLE STOTT
ASTRONAUT

NEW YORK

Seal Press
Hachette Book Group
1290 Avenue of the Americas, New York, NY 10104
www.sealpress.com
@sealpress

Printed in the United States of America

First Edition: October 2021

Published by Seal Press, an imprint of Perseus Books, LLC, a subsidiary of Hachette Book Group, Inc. The Seal Press name and logo is a trademark of the Hachette Book Group.

Print book interior design by Six Red Marbles.

Library of Congress Cataloging-in-Publication Data
Names: Stott, Nicole (Astronaut), author.
Title: Back to Earth : what life in space taught me about our home planet—and our mission to protect it / Nicole Stott.
Description: First edition. | New York : Seal Press, 2021. | Includes bibliographical references and index.
Identifiers: LCCN 2021009682 | ISBN 9781541675049 (hardcover) | ISBN 9781541675032 (ebook)
Subjects: LCSH: Stott, Nicole (Astronaut) | International Space Station—Biography. | Nature—Effect of human beings on. | Human ecology | Environmentalism. | Women astronauts—United States—Biography.
Classification: LCC GF75 .S79 2021 | DDC 304.2—dc23
LC record available at https://lccn.loc.gov/2021009682

ISBNs: 9781541675049 (hardcover), 9781541675032 (ebook)

LSC-C

Printing 1, 2021

To my son Roman.
You are my greatest blessing
(and my favorite Earthling).
I can't wait to see the good you
do in the world. You are the
reason I wrote this book.

For we Earthlings, we human beings, are the only life form that we know of that can dream, then plan and work together to achieve that dream. We are extraordinary in the universe in that our only limits are those we place on ourselves.

—Alan Bean (1932–2018), artist and Apollo Moonwalker

CONTENTS

INTRODUCTION 1

CHAPTER 1 **ACT LIKE EVERYTHING IS LOCAL (BECAUSE IT IS)** 11

CHAPTER 2 **RESPECT THE THIN BLUE LINE** 41

CHAPTER 3 **LIVE LIKE CREW, NOT LIKE A PASSENGER** 79

CHAPTER 4 **NEVER UNDERESTIMATE THE IMPORTANCE OF BUGS** 103

CHAPTER 5 **GO SLOW TO GO FAST** 139

CHAPTER 6 **STAY GROUNDED** 179

CHAPTER 7 **WHATEVER YOU DO, MAKE LIFE BETTER** 217

CONCLUSION 249

ACKNOWLEDGMENTS 265

HOW TO GET INVOLVED 273

NOTES 275

Astronaut Bill Anders became the first person to photograph our planet rising above the Moon. He took this iconic image, *Earthrise*, in 1968 during Apollo 8, the first crewed mission to orbit the Moon.

NASA

INTRODUCTION

BILL ANDERS AND HIS TWO APOLLO 8 CREWMATES, Jim Lovell and Frank Borman, were the first human beings to leave the relative safety of low Earth orbit and travel a quarter of a million miles out into space to circle the Moon. They were also the first to witness, in person, the beautiful spectacle of our colorful planet, full of life and set against the stark and seemingly endless black backdrop of space, rising in stunning contrast above the lifeless surface of the Moon.

Anders snapped a photo and became the first human to capture a photograph of this event, an iconic image now famously and appropriately known as *Earthrise*. Lucky for us, the astronauts did not keep this vision to themselves. Instead, on Christmas Eve of 1968, while still in lunar orbit, they shared it with all of us back here on Earth. This single image forever shifted humanity's perspective on who we are and our place in the universe—a simple yet powerful reminder of the undeniable reality of our ultimate interconnectivity and interdependence.

We are all Earthlings.

In 2019, we celebrated the fiftieth anniversary of the Apollo space missions that took human beings off our planet and to the Moon. I believe it's no coincidence that 2019 also marked the forty-ninth anniversary of both the first Earth Day and the formation of the US Environmental Protection Agency (EPA). *Earthrise*, the beautiful image of our home planet with its naturally perfect life support system, protected by a thin blue line of atmosphere, has been a symbol of the environmental movement ever since the Apollo 8 crew first shared it. But do we still appreciate its significance?

In 2009, forty years after I watched with my family on our black-and-white TV as the Apollo 11 astronauts walked on the Moon, I flew to space for the first time and experienced my own Earthrise moment.

As astronauts, we do our best to prepare for both the planned activities of our mission and the unexpected events that might happen along the way. Included in that preparation is communicating with our astronaut colleagues to learn as much as we can from their spaceflight experiences—including what it was like for them to see

Earth from space. For some reason, I thought I could prepare for that special space-bound experience by looking at pictures and videos. Wrong.

Let me just say that the perspective offered through the windows of the spacecraft was so overwhelmingly impressive and beautiful that it far exceeded my highest expectations. No picture, no video, and no conversation with others who had flown before could have prepared me for what I saw with my own eyes and felt with my own soul.

The view was crystal clear and glowing. The simplest way I can explain it to you is the same way I explained it to my seven-year-old son the first time I called him from space. I told him to imagine he had the brightest light bulb ever, splattered with all of the colors we know Earth to be, and that when he turned it on, it was almost too bright to look at. All the Earth's colors glowed with an iridescence and translucence I'd never seen before.

My Earthrise moment came while looking out the windows of the Space Shuttle and the International Space Station (ISS). Absorbed in that view of a translucent Earth, I realized something I'd actually known all along, something I'd probably learned in kindergarten, in 1968, the same year as the Apollo 8 mission—*we live on a planet!*

I found it strange that this simple truth could be so revelatory. But then again, how often do any of us contemplate its profundity? While it's cool that nearly everyone can see Earthrise every day, the downside is that we seem to have become desensitized to its significance. Since Bill Anders took that first photograph, *Earthrise*, images of the Earth from space have become ubiquitous. Through the wonders of technology, we see these images in lots of places every day—in advertisements, on flight tracker screens during airline travel, and through apps like Google Earth on our smartphones.

Although technology has made us more connected than ever, we somehow fail to appreciate the interconnectivity we share. We seem to forget that our interconnectivity makes all of us interdependent. Experiencing Earth as a unified whole is more than an idea for astronauts; it's a powerful reality. It is a message from the universe, intended not for space flyers alone but for all humanity. To understand the significance of Earthrise, we have to connect with these three simple lessons: we live on a planet, we are all Earthlings, and the only border that matters is the thin blue line of atmosphere that protects us all from the deadly vacuum of space.

Astronauts want to share this perspective because we believe it has the power to raise everyone's awareness of our role as crewmates here on Spaceship Earth. It is the key to peaceful and successful cooperation for all humanity, and to a future when all life on Earth not only survives but also thrives.

Now the question is: how can more people receive and take this message to heart? I believe that we each need to find our own Earthrise moment, and you don't have to go to space to do it. An Earthrise moment is any moment that stands out for you and impresses you with a sense of awe and wonder that inspires you in a life-changing way. Here on Earth we are surrounded every day by opportunities to experience life-changing awe and wonder if we're open to it. One of the simplest ways you can gain this perspective is by watching the Sun set on the western horizon and at the same time witness the Moon rise in the eastern sky. Wow! You're on a planet!

Further, we must take responsibility for our impact on the Earth—our one and only life support system—and for how we care for it. I am not alone in this sentiment: all astronauts, including the pioneers of Apollo, have expressed their own profound feelings of Earthrise awe and our deep and vital relationship with one another and our planet.

As a National Aeronautics and Space Administration (NASA) astronaut and aquanaut, I was blessed to experience our home planet from some extraordinary vantage points. Across two spaceflights, I spent over three months living and working in outer space on the International Space Station. On my first mission, I flew to the ISS on the Space Shuttle *Discovery* as part of the STS128 crew and spent three months on the ISS as part of the Expeditions 20 and 21 crews before returning home on the Space Shuttle *Atlantis* as part of the STS129. On my second mission, I was part of the STS133 crew; we flew the final flight of the Space Shuttle *Discovery*, spending two weeks on the ISS (not long enough!). In preparation for spaceflight, I spent eighteen days living underwater on the Aquarius Reef Base, an undersea habitat in the Florida Keys. It was during these experiences that the reality of our planet as our home, as our life support system, became crystal clear to me: we live on a planet, we are all Earthlings, and the only border that matters is the thin blue line of atmosphere that protects us all.

These three simple truths shine light on the interconnection of everything, on the significance of our relationships to one another and our planet, and on the need for each of us to focus on what we share in common. To embrace these truths as guiding principles would enable us to accomplish the seemingly impossible—to reverse climate change and preserve our life support system here on Earth. Most importantly, these truths drive the need for us to accept our role as crewmates here on Spaceship Earth—responsible to one another for our mutual survival on our one shared home.

Our ability to succeed in challenging undersea and outer space environments is dependent on a "here's how we can" approach.

Space travel would not be possible in a pessimistic world focused on "why we can't." Rather, space travel is based on a belief that there is a solution to even the most complex problems, and a deep respect for the possibilities opened up to us by diverse teams working together as a crew. "Here's how we can, not why we can't," is the same approach we should be taking to work together to save our planetary life support system.

The most important lessons I learned as a NASA astronaut were not about science but about people.

I felt compelled to write this book because I believe everyone should have the opportunity to benefit from what we've learned from living and working peacefully and successfully during a decades-long partnership of fifteen different countries on the ISS. It's a model for how we should be living and working together as the crew of Spaceship Earth.

As part of my research for this book, I've read many books and articles about the current state of our environment, about the interconnectivity and interdependence of all life on our planet (from the tiniest of insects to the cosmic expanse of our universe), and about the history of how we've dealt with climate change so far, the alarms that have been sounded, the looming threats to our existence, and the technology that we've misused and the technology we could use better.

The more I read, the more I became haunted by the question: how can I possibly do anything to help make life better? So much of what I read left me with a sense of gloom and doom, and with the feeling that it's too late to save life on Earth, and that it's all my fault.

Several times I've thought that I should just close my computer, stop working on this book, and instead spend more time with my family and friends outside, in nature, while it still exists. That I should hold my son close and apologize to him for the future that he'll inherit.

It's been a slippery slope. But each time I began to despair I was able to regain my senses, shouting, "Hell no!"

I refuse to believe that greed rules, or that we are incapable of choosing the right path for ourselves with respect to the long-term survival of humanity and the world we'll leave for future generations.

Granted, our history with climate change has not been a shining example of how to rally when presented with a planetary challenge. The basic science of the greenhouse effect and the understanding of our human impact on the environment have been with us for decades. We've known that the more coal, oil, and gas we burn the warmer the planet gets, and the more devastation we cause to our life support system. This isn't rocket science. Yet we've watched it happen and done little in the grand scheme of things to reverse it.

We can't change the past, but we can make a better future by knowing our own past—what we've done, what we haven't done, what we could have done differently, and how we can do better.

I'm not here just to bemoan the lack of action to this point, nor am I here to downplay the significance of the situation we find ourselves in. While this book is not intended to provide all the answers, I do believe that the success of our work on the ISS is a model of hope for all humanity. For in my research, and in my own experiences, I've found inspiring stories of individuals who have overcome seemingly insurmountable obstacles and committed their lives to making life better here on Earth, now and for future generations. The people I interviewed for this book are excellent examples of what one person can do to bend the arc of history toward our continued survival.

I've learned of initiatives and inventions that give us the ability right now to reduce our carbon footprint substantially, without having to revert to preindustrial-age living conditions.

I've witnessed what humans can accomplish together when we decide to do so. All of my work over thirty years as a NASA engineer and astronaut has been built upon finding fact-based solutions to the most challenging problems and then implementing them. At the core of our success as astronauts is acceptance of our role as crewmates as we go into each mission. The role of each crew member is important. Each of us has something to contribute to the mission of making life better on the ISS and beyond. Once we've accepted that we are crewmates, we have the obligation to become the guardians of our spaceship and all our fellow crewmates.

To behave as crewmates here on Spaceship Earth is perhaps the most valuable new skill that humanity could acquire. We just need more people to get on board.

My time on the ISS and the perspective I gained from spaceflight have greatly underscored for me the importance of my role as a crew member on Earth, where I have no more important mission than as a mom, the guardian and protector of my child. I want my son to embrace and carry with him, as he dreams of and strives to realize his future, the power that resides in the three simple truths of our reality here on Earth. My desire is that you too, as a reader, will make your own connection to these three truths: that we *do* live on a planet spinning in outer space, that we are indeed all Earthlings, and that the only border that matters is the thin blue line of atmosphere that protects us from the deadly vacuum of space.

You don't have to be a parent to understand these truths. They apply to all humans. We all are guardians of the future. As guardians, we know that our most important job is to ensure a better future for humanity's children. The way we care for our planet as our life support system goes hand in hand with our responsibility to care for our children. Every one of the people I spoke to in the preparation of this book shared this same thought in their own way. Mark Tercek, former CEO of the Nature Conservancy, shared it in the most direct way when he said, "I want to be able to look my kids in the eye and tell them I did all I could to leave the world a better place for them."

As I considered the best way to present the ISS experience as a model for our mission as crewmates together here on Earth, I decided to steer clear of offering a to-do list and to focus instead on "ways of being"—to present my take on who we need to be so that we can operate as crew on Spaceship Earth as successfully as my colleagues and I have on the ISS.

The title of each of the seven chapters is worded as a directive toward one of these ways of being. Based on some of the keys to the sustained success of the ISS program over many years, they were purposefully chosen because they are equally applicable to life on Earth. I share anecdotes from my spaceflight experience to help illuminate each directive. I provide stories that illustrate the human component of international cooperation, as well as examples of successes gleaned from the science conducted through the ISS program that have been brought back to benefit Earth. Each chapter also highlights an individual who is living as a crew member here on Earth and has leveraged their own passion, expertise, and experiences in service of improving life on our planet.

I invite you to read my book with your role as a guardian in mind. I hope that, in issuing a call to action, it provides you with guidance and inspiration for discovering how you can expand your role as a crew member on a mission to protect our planetary life support system and the future of humanity.

As seen from the Space Shuttle *Discovery*, our view of the International Space Station after the completion of the 133rd mission in NASA's Space Shuttle program in 2011.

NASA

CHAPTER 1

ACT LIKE EVERYTHING IS LOCAL (BECAUSE IT IS)

We came all this way to explore
the Moon, and the most
important thing is that we
discovered the Earth.

—Apollo 8 astronaut Bill Anders

REFLECTING ON THE FIFTY YEARS since his Apollo 8 mission, Bill Anders said that "'Earthrise'—the lingering imprint of our mission—stands sentinel. It still reminds us that distance and borders and division are merely a matter of perspective. We are all linked in a joined human enterprise; we are bound to a planet we all must share. We are all, together, stewards of this fragile treasure."[1]

Yet, in the half-century since that first Earthrise photo, our environment has deteriorated to the point where, according to the 2019 UN Report on Biodiversity, "nature is declining globally at rates unprecedented in human history." Sir Robert Watson, chair of the Intergovernmental Science-Policy Platform on Biodiversity and Ecosystem Services, warns that "the health of ecosystems on which we and all other species depend is deteriorating more rapidly than ever. We are eroding the very foundations of our economies, livelihoods, food security, health and quality of life worldwide." He adds that "the Report also tells us that it is not too late to make a difference, but only if we start now at every level from local to global."[2]

Still, we have not responded with the urgency that the situation requires. We continue to sputter, applying patchwork solutions here and there, while the environmental conditions that ensure our survival continue to degrade.

The UN report states that humans have now "significantly altered three-quarters of the land-based environment and about 66% of the marine environment of our planet," as a result of activities like corporate farming and industrial livestock production, which have increased the amount of methane released into the atmosphere, caused fertilization runoff into waterways, and led to the widespread use of pesticides. Other sources of environmental deterioration have been urban sprawl, overfishing, resource exploitation, and pollution generated by plastics and toxic chemicals. The UN report went on to say that "around one million animal and plant species are now threatened with extinction, many within decades." The rate of global extinction is estimated to be tens to hundreds of times higher now than at any previous moment in human history, and sadly, it is believed to be accelerating.

Extinction. One million species. Because of us. Let that sink in.

This threat looms for all creatures, great and small. Whether we appreciate it or not, *their* survival is essential to *our* survival.[3] Climate change, rising ocean temperatures, and dying coral reefs pose an imminent threat to our ability to breathe, as ocean plants are responsible for producing 50 to 80 percent of the oxygen in our atmosphere.[4]

Yet humans seem oblivious to the fact that what happens in my ocean is happening in your ocean. We have been so focused on maintaining man-made borders, chasing short-term gains, and demanding immediate gratification and convenience, while putting off long-term, sustainable solutions, that we have yet to embrace the most important lesson in the *Earthrise* photo—that here on Earth we are one community. We breathe the same air and are warmed by the same Sun. There is no "yours" and "mine" when it comes to the one thing that we all have in common: We all (including all animals and plants) share the same planet in space together. There's nothing that speaks more simply and honestly to everything being local than the fact that we live on one planet. Your neighborhood, and mine, is Earth.

Being an astronaut offered me many opportunities to appreciate different perspectives. In preparation for spaceflight, I completed an undersea mission in which my crewmates and I were immersed sixty feet deep in the ocean off Key Largo, Florida. That experience gave me an "inner space" perspective where the planet seemed to engulf me while from outer space I felt as if I was surrounding it. From each of these vantage points, I developed an immediate and visually stunning understanding of our interconnectivity—that all that connects us defines our relationship to and our interdependence upon one another. This understanding of our interconnectivity has become an integral part of who I am and how I feel about my relationship to everyone and everything on Earth—and to the Earth itself. I believe it will remain a powerful influence on all my actions and choices for the rest of my life.

Not only was the opportunity to be immersed in our planet deep beneath the surface of the ocean the absolute best analog to what life would be like on a space station, but it also opened up for me a whole new understanding of this planet we share. There is so much life down there that we're not aware of in our day-to-day lives.

As a crew of six, we spent eighteen days on a mission in the undersea habitat called Aquarius.[5] This school bus–sized habitat that sits on the ocean floor is about the same size as a single International Space Station module, and it's the only undersea habitat and laboratory of its kind in the world. Unlike the ISS, all the critical elements we needed to survive (for example, air, electricity, communications) were supplied to us through cables and hoses from a life support buoy on the surface. For us as a crew, however, everything about the experience of life on Aquarius paralleled the way we would live and work on the ISS.

After an hour at depth, our bodies were saturated with nitrogen, so there was no zipping up to the surface in the case of an emergency because a quick "escape" to the surface would mean serious injury, likely death, from the bends. All our mission activities on or in the vicinity of Aquarius were analogs to the kind of work we would do in space, including the science experiments that we performed and some that were performed on us, and scuba dives to test out surface exploration techniques for future use on the Moon or Mars that would emulate a spacewalk. As we would do on the ISS, we lived and worked in isolation in an extreme environment that required us to deal with emergencies as a crew in order to ensure our survival, and we enjoyed a view out the window that was overwhelmingly impressive and difficult to describe.

On the eighteenth and final day of our mission, as we prepared to surface, I sat at the galley porthole with my dear friend and fellow astronaut Ron Garan. As we gazed out the window, we reflected in awe at the time we'd just spent as part of this undersea world, and both of us agreed that even if we never got to space, we already had been blessed to have experienced our planet from this perspective. We never had the chance to fly together in space, but we often talk about the undersea adventure we shared.

While the extreme environments of undersea and outer space provided me with the opportunity to experience our planet with a whole new sense of awe and wonder, they also allowed me to appreciate the very basic challenge associated with both—survival. To live in either of these places is a difficult proposition. Whether scuba diving for recreation or spending eighteen days doing research and training in an undersea habitat, you need special equipment to survive. The same is true for outer space. The experience made me appreciate how perfectly designed our planet is to take care of all this for us. We do our best with our undersea habitats and our spaceships to mimic what Earth does for us naturally. Earth still does it best.

The fact that a planet is our shared home in space became real for me while I was living on a spaceship. I became keenly aware that regardless of whether I was on Earth or on a spaceship, the one thing these two places have in common is that they're both our life support system *in space*.

Granted, it has been a relatively short period of time since 1961, when people first left our Earth-bound home in space to establish homes in outer space like the International Space Station, and in that time very few people have had the opportunity to make that trip beyond the protective atmosphere of our planet. (Of the approximately 108 billion people who have ever lived, fewer than 600 have flown in space.[6]) I consider myself blessed to be one of the few who have. Yet, whether or not we've lived on a space station, we all must consider how we can access this perspective of awe and appreciation of our planet's unique ability to provide a safe, comfortable home. It's key to our survival.

The awesomeness of this vantage point is not lost on astronauts, as most of us spend our free time gazing out a spaceship window. Everyone's favorite pastime is taking pictures of the Earth. If you ever do travel to outer space, even if you weren't a photographer before, you'd become one. Orbiting Earth at speeds of 17,500 miles per hour, or five miles per second, I had to develop a steady hand to capture what I saw below. If I didn't get a particular shot on the first try, I'd have to wait at least another ninety minutes before I'd have a chance to see it again.

There is no better example of an astronaut photographer than my friend Don Pettit. A "MacGyver" in the astronaut world, there's not much that Don doesn't know how to jerry-rig or fix. He demonstrated this skill through his photography while he was in space, and he has shared his expertise and stunning images with the rest of us. Two of Don's favorite subjects to photograph were city lights on the planet at night and star trails.

To capture city lights takes an especially steady hand, given the speed at which we're traveling. During Don's first spaceflight (ISS Expedition 6, November 2002 to May 2003), he solved the problem of clearly tracking and photographing the city lights below by "MacGyver-ing" a Makita drill into a tracking mount for the camera. In his own words,

> I assembled a "barn door tracker." It's based on the fine gimbal movements in the IMAX camera mount for the Destiny Lab window. I figured out a way to mount a threaded screw and nut (scavenged from a Progress rocket) and drive it with a Makita drill driver....All of these modifications clamp on to the IMAX mount and do not change its original function in any way....I manually compensate for the station's motion by looking through the spotting scope and running the drill at the same time. It takes a bit of practice, but you do learn to track.[7]

Don flew a total of three space missions, and each time he found a way to innovate and creatively photograph the Earth below and the stars in space that surround us. His star trail images have become widely popular for their beautiful presentation of the universe, and he left all other astronauts thankful that he was so willing to share his tips for taking these incredible shots. While not nearly as technical as Don's, my own photography in space improved over time and allowed me to do a pretty good job of capturing images I could keep as memories and share with others once I was back on

Earth. From space, the view of Earth is like a work of art—so beautiful that it also inspired the watercolor painting I created there.

Other crew members enjoyed their favorite Earthly pastimes as well. They sewed, played musical instruments, and even experimented with cooking. During downtimes, we called family and friends. We also participated in some of our earthly duties. During my first mission, my American crewmate Jeff Williams and I voted. I even received a jury duty summons, forwarded to me by my husband via email. I replied that I would be unable to attend owing to the fact that I was in outer space. I was excused.

I spent my two spaceflights primarily on board the International Space Station; of my 104 days in space, eight were spent traveling on the Space Shuttle to and from the ISS.

Everything about a spaceship has to take into account the survival of the crew—the humans who will live and work on it. Design teams must consider how those living on board will have clean air to breathe and water to drink, how they will be protected from radiation from outer space, and how they will maintain the air pressure necessary to keep their bodies intact. They must also consider what kind of food crew members will eat and how the food will be prepared; how they will keep their bodies strong in microgravity; how the temperature will be controlled; how they will communicate with mission control and their families on the ground; how their waste will be collected and disposed of; and how emergency situations will be managed.

There are so many challenges to overcome just to be able to survive, let alone the challenges associated with implementing the science experiments or the politics associated with a multinational program like the ISS. And yet, since the year 2000, and with the participation of fifteen countries represented by five international space agencies, the ISS program has operated peacefully and successfully with up to seven international crew members in space and tens of thousands of mission team members here on Earth.

The size of a football field, the ISS is a self-contained, orbiting habitat and laboratory. International crews live and work together in the habitable portion of the station, which is roughly the size of a six-bedroom house. It's spacious when you consider that you're living and working in space, but not so commodious when you realize that for three months it's your entire world. There's no walking out the door for a stroll, or casually hopping in your own spaceship to head home at the end of the workday.

One of the key differences between living on a spaceship like the ISS and living on Spaceship Earth is gravity. Gravity is integral to how our planet supports life. It's another one of those things we take for granted, but it's pretty compelling to think that we wouldn't have an atmosphere or oceans without it. It's also why reduced gravity poses such a challenge to our survival in outer space. On the ISS, we operate in an environment of "microgravity," which is sometimes referred to as "zero gravity." When we go back to the Moon and later to Mars, we will be operating in reduced-gravity environments (one-sixth and one-third, respectively, as compared to Earth); during trips there and back, we'll likely encounter a mix of microgravity and artificial gravity. (Artificial gravity is taking an action that creates a force that feels like gravity. One way to do this is by rotating the spaceship.)

Gravity is a force that tries to pull two objects toward each other. No matter where you are in space, there is at least a minuscule (micro) level of gravity. Gravity is also the force that keeps all the planets in orbit around the Sun, and here on Earth it's what keeps you on the ground and what causes objects to fall. Microgravity, like what we experience on the space station, is a condition of virtual weightlessness or continuous free fall, which has always seemed so mysterious to me. The way I like to think about it is to compare it to throwing a baseball. If you throw a baseball with just a little force, it will drop right in front of you, but if you give it all you've got, it will go really far before it comes back down. The same happens to us on the space station, as we're essentially in a continuous fall around the planet. We've put enough thrust into getting the space station to orbit around the Earth, so that we just keep falling around it instead of falling back into it. The same thing happens for a quick second when you're going over the top of a roller coaster, except in space we just keep floating.

To be able to float and fly is one of the unique things about living in space, and the most fun, but it also brings some of the greatest challenges. When I first got to space, I was a little clumsy as I tried to get used to life in three dimensions: I'd hold on too tightly to things, push off too hard, and try to reconcile some kind of up and down that didn't exist. But our brains and bodies are incredible. We figure out very quickly how to adapt to new environments, even extreme ones. In some ways, this ability is very helpful, but in others it's not so good. The body figures out how to move gracefully in three dimensions, how to float and move smoothly and with a gentle touch, how to hang on by one toe instead of tightly grasping with a hand, how to take advantage of the liberating feeling of floating and flying. Our spines elongate; I "grew" an inch and a half while in space. (To the dismay of shorter astronauts, we've discovered that you shrink right back to normal when you get back to Earth.) Those are the good things.

The not-so-good things are that our brain and body figure out that we don't need the same bone density or muscle mass to survive in the microgravity environment, so they expend less energy to maintain them. The result is rapid bone and muscle loss (including the heart muscle). Another physical challenge is that body fluids shift toward the head; you'll notice that astronauts' faces look fuller in space than on Earth. While a microgravity experience is a great overall "lift" for your body, it also has some troublesome physiological impacts. Increased intracranial pressure, for instance, is believed by the medical community to be the likely cause of the degradation of vision that many astronauts experience while in space; some do not recover their normal vision after returning home.[8]

Another risk with potentially serious implications for human spaceflight was discovered during an ultrasound study in 2019 that took place on board the ISS: the blood flow in the jugular veins of eleven astronauts was evaluated preflight, at different intervals during their flight, and after their return to Earth. The onboard findings indicated a surprising flow reversal in six of the eleven astronauts and the formation of a small blood clot in one, which was especially concerning. In space just as on Earth, a blood clot increases the risk of cardiac arrest and stroke. The crew member with the clot was treated with anticoagulants for the remainder of the flight, the clot shrank before returning to Earth, and it disappeared within ten days after landing. To protect the identity of the astronaut participants, the specific dates of the onboard study have not been released, but the results of the overall study were released in 2019.[9] Some of the other negative side effects associated with microgravity include weakened immune systems, kidneys stressed by bone loss, cell behavior changes, and slower wound healing. As we spend more time in space, we continue to find out more about how microgravity affects our bodies.

It's important for us to understand and counteract as many of these effects as we can, so we can return to Earth at the end of our missions and continue to live healthy lives. We address these effects in space through countermeasures like exercising two hours each day, with a mix of resistive and aerobic training, to counteract the bone and muscle loss. As a point of reference, in microgravity bone loss occurs at a rate of 1 to 1.5 percent a month, about twelve times the rate of bone loss that occurs in elderly men and women on Earth, leading to an acceleration of age-related changes similar to osteoporosis.[10] So by living on the ISS, in addition to learning how to manage bone loss for astronauts in space, we can learn a lot about the process and develop ways to help prevent and treat osteoporosis here on Earth.

The research performed in the microgravity environment of the space station is important because it allows us to take gravity out of the equation, thus providing us with the opportunity to learn all kinds of new things about stuff we thought we already knew a lot about. Pretty much everything behaves differently in microgravity. Fire burns in a different shape—flames are rounder—and we can better understand how different fuel mixtures burn cleaner or more efficiently. Protein crystals, which are used by scientists to study the structure of proteins for medicinal uses, are easier to study because they grow larger and form more perfectly in microgravity. We are continually learning more about the impact of the microgravity environment on our bodies, but from a human-in-space standpoint, the ability to float and fly (and even paint with watercolors) is a wonderful part of the space adventure.

Having had the opportunity to live and work as an astronaut for an extended time in space, I now get great joy from following the progress of other missions through the NASA TV station, website, and social media, where I can watch other astronauts flying their missions and working on the ISS. I'll always remember what it was like to watch astronauts in space before I flew for the first time, and how I wondered what it would feel like to float and fly in three dimensions. Now I watch my friends in space with a total understanding of the experience. It seems normal and natural. I also love to keep track of the different kinds of research being performed. I marvel at how crew members have become more like lab technicians, doing hands-on work and engaging in the actual science. And I remain in awe of the successful international partnership that makes it all possible and has endured for the benefit of life on Earth.

It's especially exciting to watch someone I know flying her first mission to space. In 2018, my friend Dr. Serena Auñón-Chancellor spent 197 days on the ISS.

Serena and I first met when I was in Star City, Russia, preparing for my first spaceflight, and she was the resident flight surgeon for NASA astronauts training there. (A flight surgeon is a physician who specializes in aerospace medicine, the medicine associated with people who fly professionally, including astronauts.) Astronaut preparation for a long-duration spaceflight on the space station involves several years of mission-specific training, and at least 50 percent of our time is spent outside the United States with our partners at their facilities around the world. Outside Houston, Texas, US astronauts do the majority of our training time in Star City, located about thirty miles outside Moscow. The training center there, the Yuri Gagarin Cosmonaut Training Center (GCTC), is named after the first human to fly in space. (On April 12, 1961, Soviet cosmonaut Gagarin completed one orbit of the Earth.) All cosmonauts (the Russian term for "astronauts") have trained there, as well as all international astronauts flying to the ISS. Some of the instructors who trained Gagarin were still working at the GCTC in 2020. It's a historic place, with sculptures, statues, stained glass, metalwork, and other artistic tributes to the significant spaceflight events that have taken place over the years. It was an honor for me to train there alongside my cosmonaut and astronaut crewmates from around the world, and to get to know the people who live and work there, many of whom I know will be my friends for the rest of my life.

Each training session in Star City was usually four weeks long. The three years I spent training for my station flight went something like this: four weeks in Star City, four weeks home, two weeks in Germany, four weeks home, four weeks in Star City, four weeks home... Tokyo...Montreal...four weeks home, four weeks Star City.... So inevitably I spent some holidays away from my family.

One of my fondest holiday memories includes Serena. She and I and a group of international astronauts and training personnel from several different nations and walks of life ended up spending Thanksgiving in Russia. Shopping at the US embassy store in search of the crunchy onions for green bean casserole, Serena and I watched a woman take the last can from the shelf. We locked eyes and joked that we could take her...but decided not to. We next went to the local street market and found carrots in giant tubs of the blackest dirt we'd ever seen. The vendor pulled them up from the dirt, and it took a lot of scrubbing when we got back to our cottage to reveal the beautiful orange beneath (they were delicious); we also purchased an outrageous number of potatoes, which resulted in what seemed like the largest tub o' mashed potatoes ever served at a Thanksgiving meal. When we discovered that the turkey was way too big for the oven in our cottage, Serena and I had to rerig the oven rack to make it work.

Even though we went a little overboard, we wanted the celebration to feel like it would have at home, with all the traditional fixin's, and we wanted our international friends to experience it as they would if they were guests in our homes. Everyone wanted to contribute to the meal, so we had quite the assortment of the standard Thanksgiving fare along with some delicious Russian dishes too.

At one point during the evening, I looked around and was amazed at all these incredible people from so many different places around the world who had somehow come together in Star City as part of the space program, and I was thankful.

The most difficult thing to me about astronaut training was learning to speak the Russian language. The official language on the space station is English (I'm thankful for this as well), but everyone also has to speak Russian because the Russian Soyuz spacecraft has always been a rescue vehicle, and everything about its operations is in Russian, from the procedures and instrument panels to the communication with the mission controllers on the ground. One of the things I'm most proud of is getting to the point where I could actually communicate in Russian in the way that was required in support of the mission. (Ginormous thanks to my very patient language instructor, and dear friend, Waclaw Mucha!) Still, I worried whether what came out of my mouth actually meant what I intended. That Thanksgiving in Star City, I found the courage to raise a toast in Russian at the start of the meal. While I did keep it simple, I tried to convey thanks to the people in the room for the amazing things they were making happen in space, all of which was for the greater good of everyone here on Earth, and to thank them for the friendship that had created this big international family we all shared.

Because I was speaking in Russian, my friend and fellow astronaut Mike Fincke provided English translation. I was so pleased that he was able to understand and translate what I was saying (and that it actually came out as I intended). There is no doubt that the work and science we're doing together in space matters, but to me, the way we're doing it—as a community of humans from Earth—is more important.

After the ISS had been in operation for about fourteen years, one of my dear friends and fellow astronaut classmates, Kevin Ford, coined the phrase "Off the Earth, For the Earth," as the slogan for his own crew's mission, but because it so perfectly expresses the overall ISS mission, it has been the motto of the International Space Station program ever since. It is beautifully appropriate. You'll read more than once in this book how all the work going on with the ISS is ultimately about improving life on Earth.

Some of the questions about that work I get most often are

- Why space?
- Why can't we do that science here on Earth?
- What has the work on the space station done for me lately?
- Why are we spending so much money in space?

These are all reasonable questions, and ones I think should be answered, in particular the question about how the work in space benefits our life support system here on Earth. (And my answer to "why are we spending so much money in space?" is that no money is spent in space—all the money spent on space exploration is spent right here on Earth.)

November 2, 2020, was the twentieth anniversary of continuous human presence aboard the ISS, which by then had hosted 239 people and more than 2,700 science experiments. The research on the space station is broken down into six broad areas of study: biology and biotechnology, earth and space science, human research, physical science, technology, and education.[11] I've found that an easier way to get our arms around the ISS research is to think of it from either a human (personal) health perspective or a planetary health perspective.

I was pleased to have the chance to interview Serena Auñón-Chancellor for this book. Not only a friend and colleague, she is also an astronaut, medical doctor, and life science guru who spent six months on the ISS. I posed my questions to her from the standpoint of personal and planetary health. We spoke via Skype, I from my home studio in St. Petersburg, Florida, and Serena from her office in Baton Rouge, Louisiana, where she's taken a faculty position at the Louisiana State University (LSU) Health Sciences Center. Not one to ever shy away from applying her skills and expertise where they're needed, she was practicing medicine in the hospital, teaching at LSU, and still supporting medical-related issues for the astronaut office. (In particular, she was responsible for all COVID-19 protocols for the astronaut office.)

"I'm happy because what I'm doing right now is working in the three areas I love: medicine, aerospace medicine, and the astronaut office," Serena said. "If someone were to ask me what's in store for the future, I would just say, 'Great things.' I don't know what those are. They tend to kind of pop up out of nowhere; opportunities just come. I try to step through that door when I see them. I know there's a lot coming down the road, but I just love being on Earth right now."

I was happy Serena was on Earth too, so I could speak with her, and it was nice to see her on my computer screen. With our wavy dark hair and dark eyes, we look a lot alike, so much so that when we meet I feel like I'm talking to one of my sisters. We started our conversation with the ISS motto "Off the Earth, For the Earth." She agreed that it is perfect for describing how we live and work together on the ISS.

"It's interesting, though, how so many people really don't understand how the 'For the Earth' part is being realized." Then Serena shared something that even I hadn't known. "While most of the science overall that's being done has benefit to both life on Earth and further exploration of space, in the area of life science at least 70 percent of it is being performed solely for the benefit of human health back on Earth."

NASA and our international partner space agencies have always struggled with how best to communicate the amazing things happening every day in space. There has been significant improvement in this communication over the past few years, as social media and other creative mechanisms have helped NASA reach audiences that might not otherwise know that an international space station even exists, let alone anything about the work that's being done there or by whom.

One of the many things that Serena did well while she was in space was to share information with the general public through YouTube videos about the science she was performing, and why it was so important.

One video she made for an experiment called "Angiex Cancer Therapy" stood out for me.[12] This is research designed by a company named Angiex to study the growth of endothelial cells (the cells that line blood vessels and transport blood through the body) and the effectiveness of cancer treatment drugs in stopping the supply of blood to cancer cells and tumors. Serena explained to me that individual cells behave in the microgravity environment of space in much the same way they behave naturally inside our fluid-filled bodies.

"The cells in our body aren't flat, they are three-dimensional, and they grow and float and move in a suspended state in our bodies.

Similar to how we float and move through our spaceship in microgravity," she said.

"When we do experiments on Earth that take cells out of their naturally suspended state and try to study them in a flat plate, then they're impacted by gravity, and they're stuck, and their entire structure changes. In space, we can provide the cells with an environment similar to the inside of the human body, so they grow better, and it's a much better place for studying their behavior."

Of course, we also spoke about how awesome it is for us to float and fly in space and the very special experience of being there and the life-changing perspective of seeing our home planet from space. Serena chuckled during this serious and philosophical conversation because it brought to mind another memory she had of working with Angiex. She explained to me how special the cells in these experiments are.

"They are the tiniest form of life—they are like little crew members living with us." It gave me goose bumps to hear her describe them this way. She went on to say that she was remembering the first time she put these cells away in their storage compartment.

Smiling, she said, "I had just finished a media exchange [feeding the cells], and the ground [mission control] told me it was time to put them back in their stowage rack in the lab. I remember floating up to the rack and opening the door to the locker, and as I was pushing the cell cartridge into the drawer, I could feel the heat, and it just felt so nice and just so comforting. I was still on headset with mission control, and I said, 'Oh my gosh, it feels so nice in here.'

"And they replied, 'That's because it's body temperature, so it will feel like they're home.'

"And I said, 'It does. It feels like home.' Everybody was laughing at me on the ground, but it just felt so good."

Serena shared more about the life science (personal health) research she supported on station—how impressed she was to be able to see in person how those endothelial cells and protein crystals really do grow better in space, and to learn how taking gravity out of the equation provides an opportunity to better understand disease processes. By the way, major pharmaceutical companies like Merck, Eli Lilly, Novartis, and Angiex are sponsoring experiments at the space station because this better understanding will support the development of more effective pharmaceuticals.

A small start-up company called LamdaVision, based in Connecticut, is working on thin, protein-based membranes that get implanted behind the retina to replace the function of damaged cells. As explained by Dr. Mike Roberts, deputy chief scientist at the Center for the Advancement of Science in Space (CASIS), which is responsible for the management of the ISS National Laboratory, these retinal implants will help macular degeneration or retinitis pigmentosa patients recover their vision. As reported in *BioIT World*, a "more stable and homogenous product" can be manufactured in space than could ever be produced on Earth.[13]

Organizations like the Michael J. Fox Foundation are taking advantage of the ISS microgravity environment to study the LRRK2 protein, which is associated with the development of Parkinson's disease. "Studying those crystals in space is actually hugely beneficial for scientists back on Earth," Serena said, "since the extreme low-gravity environment actually allows the crystals to grow larger, revealing important intricacies of how the crystals are structured, which is key to devising a strategy to inhibit them. This is being done with the hope that it could help advance the development of drugs to treat Parkinson's."[14]

Serena and I recalled how amazing it is to be on the ISS as an astronaut and to be able to support all kinds of research for scientists on the ground. Often, we had no expertise in a particular research area, so we would train and learn to act as the implementer. We became the hands and eyes and ears of the scientists on the ground who had designed the experiments and trusted us to perform them. We took that role very seriously. For most of these scientists, the research we conducted on the ISS represented their life's work, and possibly their only chance to collect needed data.

We also spoke about what good "subjects" we ourselves made for the research. Everything about our bodies changes in some way when we live in a microgravity environment in space, and almost all of these changes can be studied with respect to some parallel feature of our bodies here on Earth. So throughout our missions, we were required to collect and sample pretty much every output product—blood, urine, spit, skin, hair, and so on—and then prepare it to be frozen and sent to the ground for study. Astronauts in space make a wonderful surrogate population for studies on the impact of phenomena like bone and muscle loss, radiation exposure, immune response, aging, vision, and even cognitive changes due to the way our bodies respond to the microgravity environment.

In the area of biomedical research and human health, scientists appreciate that disease is influenced by both genetics and environment. On Earth we have discovered ways to begin unlocking the genetic part of the equation through the work of programs like the Human Genome Project; in space, scientists have access to dramatic environmental change, so they can study the influence of microgravity, intense radiation, and other aspects of the extreme environment on our genes.

Partnerships between NASA and organizations like the National Institutes of Health (NIH) have been in place since the earliest days of the human spaceflight program, in recognition of the value of the space environment for research and of the parallels and synergies of what we can learn in space to make life better on Earth. NASA's partnership with the NIH has been reinvigorated during the ISS program, and it continues as both agencies share innovative ideas for addressing scientific and technological challenges through cooperative research that allows us to expand the scope of human exploration in space while improving life on Earth.

Echoing Serena's description of the benefits of microgravity research, Dr. Christopher Austin explained in a *Scientific American* blog about the NIH relationship with NASA that "a lot of the things that slowly affect humans on Earth affect astronauts much more rapidly." Dr. Austin is the director of the National Center for Advancing Translational Sciences (NCATS) at the National Institutes of Health. He's also a developmental neurogeneticist, and he has served as the liaison to NASA for the US Department of Health and Human Services (HHS) and the NIH. "That's a real problem for NASA, but it's a real opportunity for us... to study these disorders on a time scale that would take years and years on Earth." He went on to observe that "humans have long looked to the sky and dreamed about what lies beyond. Today, scientists are looking to that same sky in hopes of understanding what ails us on Earth."

In a recent sci-fi-like approach to biomedical research on the ISS, a series of experiments called "Tissue Chips in Space" have been initiated through this NASA-NIH partnership. In many cases, the astronaut still makes the best research subject, but just as often, if not more commonly, it's not safe for the astronaut to serve in that role. In those cases, the scientists must find alternative models or subjects, and they often resort to animal subjects, who are typically poor alternatives to human ones. "Tissue chips" provide an excellent and humane alternative, and that comes as a great relief to me, as I have never been a supporter of using animals for research.

Tissue chips, while still in the early research stages, are a cutting-edge system of miniature 3D models of real human tissues, such as heart and kidney cells, that are about the size of a thumb drive; they are designed to mimic functions of the human body and can be tested to see how they respond to stresses, drugs, and even genetic changes. According to the scientists, tissue chips function like a bunch of tiny astronauts they can study in space. Exposing the tissue chips to microgravity provides additional research benefits because the changes to human cells in microgravity resemble an accelerated version of the aging and disease processes, so the scientists can make observations over the course of a few weeks that might take years in a laboratory on Earth.

In addition to the research being performed on the ISS associated with our personal health in space and on Earth, ongoing scientific research is conducted there to study the health of our planet and the condition of our planetary life support system.

Serena told me that one of the keys she's found for communicating the importance of the work being done on the space station is to make it personal—to establish relationships important enough that the audience will care. Clean air to breathe and clean water to drink are personal concerns for everyone, so it works well to explain how the research we're doing on the space station—and even how the station itself works to provide clean air and water for us in space—can help ensure that these resources continue to be available here on Earth.

During her time on the ISS, Serena worked on two experiments associated with algae production, one called "Space Algae" and the other called "Photobioreactor." The "Space Algae" experiment explores the genetic basis for the productivity of algae cultivated in space and whether this requires genetic adaptations or not. Algae may perceive microgravity as a physical stress, which can trigger production of high-value compounds (for example, oxygen) that could be used to produce pharmaceuticals and other health-related products. The "Photobioreactor" is a small-scale, closed-loop system being used to demonstrate how microalgae can efficiently convert water, light, carbon dioxide, and methane from the ISS environment for the sustainable production of oxygen and biomass (food). Both of these experiments have equally useful applications in space and here on Earth.

Mounted to the exterior structure of the ISS is a science instrument called the Hyperspectral Imager for the Coastal Ocean (HICO).[15] With its altitude and the inclination of its orbit, the space station crosses about 80 percent of the Earth's surface as it travels around the planet, including all tropical and most temperate coastal regions, where harmful algae blooms have become a major threat. This makes the ISS an ideal platform for an instrument like HICO and for studies of harmful coastal algae blooms like "red tide" (something those of us in Florida are too familiar with). Other sensors on board the station have been used to monitor vegetation, ocean surface conditions, floods, coral reefs, weather systems and climate patterns, environmental impacts of natural and unnatural disasters, carbon emissions, and urban growth, addressing a wide range of humanitarian, environmental, and commercial concerns.

As I follow along with the science happening on board the ISS and its value for improving life here on Earth, I recall Serena's comment about the best way to communicate this value. Instruments like HICO are monitoring environmental conditions that exist in my own backyard, but also anywhere on the planet that has warm coastal water. What's happening in my backyard is happening in backyards all over the planet. What I experience as local is actually global: it is happening on a planetary scale.

The "Off the Earth, For the Earth" value of the ISS is not limited to scientific results or successful international relationships—it is also related to the way we have built the space station to overcome the challenges of the humans living there. The same mechanical systems that we use to support life on the ISS have also allowed us to think differently about how we can utilize the same technology to our benefit here on Earth. The systems developed for solar power generation, water production and purification, air quality management, and communication in space are all examples of technologies that have been brought back for the benefit of life on Earth.

From the vantage point of the International Space Station, we clearly understand how the inventor and visionary Buckminster Fuller came to promote the metaphor of "Spaceship Earth." In 1968—the year of Apollo 8 and *Earthrise*—Fuller's book *Operating Manual for Spaceship Earth* was published. While the idea of Earth as a spaceship dates back to 1879, when Henry George wrote in *Progress and Poverty* that "it is a well-provisioned ship, this on which we sail through space," others have since referenced it in this way as well, including US ambassador Adlai Stevenson who in a 1965 UN speech said,

> We travel together, passengers on a little spaceship, dependent on its vulnerable reserve of air and soil; all committed for our safety to its security and peace; preserved from annihilation only by the care, the work, and, I will say, the love we give our fragile craft. We cannot maintain it half fortunate, half miserable, half confident, half despairing, half slave—to the ancient enemies of man—half free in the liberation of resources undreamed of until this day. No craft, no

crew can travel safely with such vast contradictions. On their resolution depends the survival of us all.[16]

To me, though, no one pulled this idea together better than Fuller. To this day, his work may be the most comprehensive account of Earth as a spaceship flying through space with a finite number of resources. In his words, "I've often heard people say, 'I wonder what it would feel like to be on board a spaceship,' and the answer is very simple. What *does* it *feel* like? That's all we have ever experienced. We are all astronauts... aboard a fantastically real spaceship—our spherical Spaceship Earth."[17]

We travel on a planet that spins at one thousand miles per hour to make one full rotation in our twenty-four-hour day while at the same time Earth is orbiting the Sun at an almost unimaginable speed of sixty-seven thousand miles per hour, yet we simply perceive it as home.

This idea of Spaceship Earth has always intrigued me, but it was brought to life during my time living on the Space Shuttle and the ISS. Living off our planet poses a complex challenge. Everything that allows us to survive on a spaceship is created through mechanical systems that depend upon the passengers' ability to work together as its crew. I would argue that our planet is one of the most amazing natural mysteries, but we might be very well served to start thinking of Earth as a mechanical vehicle too—with the crucial difference that Earth, unlike a vehicle, is wired to thrive. Perhaps more than maintenance, our planet requires stewardship.

Earth's life support system consists of four main interconnected subsystems: the atmosphere (air), the hydrosphere (water), the geosphere (rock, soil, and sediment), and the biosphere (living things). Life on Earth depends on three interconnected factors: energy from the sun; the cycling of matter, or the nutrients needed for survival; and gravity, which allows the planet to hold its atmosphere and causes the downward movement of chemicals in the cycles of matter.[18]

A spaceship mimics Earth in that it must operate as an integrated life support system. The main components of a spaceship's life support system parallel Earth's natural systems by necessity; on the space station, we call it the Environmental Control and Life Support System (ECLSS—pronounced E-kliss). There are fancy mechanical subsystems that create, monitor, and keep our air and water clean; that maintain and monitor atmospheric pressure, temperature, and humidity; and that even help detect and suppress fires. Energy for all this is generated from the Sun, and all this lifesaving atmosphere is held in place solely by a thin metal hull.

In our normal daily lives here on Earth, few of us consider where the clean air we breathe or the water we drink comes from. It's even less likely that any of us are thinking of how critical to our survival is the protective atmosphere that's wrapped around us. And even fewer of us might ever be aware that everything we need to survive is interconnected. This really hit home for me while I was living in space.

I was mesmerized by seeing the Earth from space, and whenever possible, I floated to a window to be surprised yet again by the view. I was especially captivated by the views from the window as we passed over Earth where it was night. The glinting lights below outlined where people lived, in contrast to the deep darkness of the oceans that cover most of our planet. The ever-changing weather was especially dramatic. The lightning of a thunderstorm in Florida whipped its way around, flashing light over Earth like neurons firing across a brain.

When I was a child growing up in Florida, I had imagined that thunderstorms were happening only over my town, and that when they were gone, they were gone. It had never occurred to me that the storm was zooming around the planet.

From space, I saw that lightning never exists in one place. It's constantly on the move. This revelation led me to see firsthand the life-changing truth that, whether we're aware of it or not, whatever happens on one part of the planet affects the whole.

One day very early in my first spaceflight, I thought about how separated I was from Earth, and farther away than I might ever be again. However, in contrast to the physical separation, I felt more connected to everyone and everything below than I ever had when I was living right in the middle of it all. How could I be so distant yet feel so close? Then it occurred to me. It wasn't just about the profound impact of seeing Earth out the window.

My time on the ISS enabled me to develop connections with people from several different nations, some of which were involved in serious conflicts back on Earth. Yet on the ISS, we all got along just fine. Even though we didn't always agree on everything, the common purpose of our mission gave us the basis for resolving disagreements. Living in space with my multinational crewmates in a self-contained environment taught me that if we can peacefully and successfully live together on a space station, with all its complexities and challenges, then we absolutely can live together peacefully and successfully on Spaceship Earth. The connections and friendships we formed during our time together made each of us realize how much we all had in common. It helped us acknowledge how close we were to one another, not just in space, but also back home on our planet. We realized there is no *us* and *them*; there are no borders—these are all man-made illusions. In truth, whether on the ISS or on Earth, it is just us, all of us, traveling through space together.

Most people today figuratively hold the whole world in their hands through that extra appendage, the mobile phone. It has given us access to what seems like unlimited information. Physical distance between people matters less now than ever before because of the information we can access about one another. We know much more about the conditions and challenges that people face in all different areas around the planet. The data and information available to us now, literally at the tips of our fingers, should make it impossible to ignore the fact that not everyone has access to the same resources or possesses the same ability to thrive.

This plethora of information packaged so neatly in the palm of our hand should be reinforcing the reality that we are all in this together, that we all share this same planet in space as home, that global is local and local is global—everything is local. Yet, regardless of how much connectivity we have, we still tend to think of who we are as separate from where our problems lie.

With all that information and connectivity available to us, I believe that by raising awareness of the very positive things that are happening on the ISS as it circles our planet every ninety minutes and that are also taking place here on Earth, we can bridge the gap between connectivity and interconnectivity. The willingness to acknowledge our interconnectivity allows us to experience our own Earthrise moment—to understand the uniqueness of our habitat here on Earth, our bodies' unique attunement to life on Earth, and the urgency of the need for all of us to work as a crew to survive on Spaceship Earth. Because everything *is* local. And everything we do affects everyone else, everywhere else. Politically, technologically, socially, environmentally, you name it—everything is connected. Every. Single. Thing.

Floating in the windows of the International Space Station Cupola Module, with Earth in the background. This is every astronaut's favorite place on the space station.

NASA

CHAPTER 2

RESPECT THE THIN BLUE LINE

AS WAS MY NIGHTLY RITUAL ON THE ISS before going to bed, I floated in front of the window by myself to enjoy the view. On one special night, while it was also night on Earth, I watched the lights twinkle and the clouds move below, searching for a glimpse of the stunning flashes of lightning I sometimes saw or maybe some colorful curtains of aurora. Then, out of nowhere, I was startled to see a streak of light zoom by below me, between my spaceship and Earth toward the glowing thin blue line of atmosphere that wraps around the planet.

What the heck was that? I thought.

I flew down to another part of the station to find one of my crewmates, Mike Barratt, who was still awake. I asked him about it, and he said, very nonchalantly, "Oh, that was a shooting star."

As I turned to fly back to the window, I said, "Well, it would have been nice to know I might see something like that!"

A shooting star below me. I was shocked because it didn't make immediate sense to me. I had to wrap my mind around the idea of seeing a shooting star *below* me. All my life, I'd only been aware of seeing one, with any luck, above me while looking up from Earth *toward* space. The shooting star was so beautiful and surprising. I floated in front of the window a while longer, hoping to see another one, but no such luck. Then I realized, though, another stroke of luck, and thought: *Wow, I'm glad I* saw *it because that means it didn't hit the thin metal hull of my spaceship.*

Viewed from the Earth's surface, the sky seems to go on forever. Any day when you walk outside, whether the sun is shining or it's cloudy and raining, or you're out at night and all you see is darkness or the stars shining above, if you take a moment to look up and consider the sky, it can seem limitless, as if there is no end to the air that wraps around us. The truth is that our atmosphere is a mere sixty-two miles *thin.* To get a sense of just *how* thin this is, imagine you're on a family trip to Florida, and you're making the seventy-mile drive from the Kennedy Space Center to Disney World. If you were to take that distance and drive it straight up, you'd reach the edge of our atmosphere before you'd arrive at the Magic Kingdom. While an analogy like this can be helpful, I have found no better way to understand this *thinness* than the view we get of Earth from space.

We take for granted that the atmosphere is there and doing its job for us—blocking the Sun's dangerous radiation, keeping the temperature in a comfortable range, and containing breathable air with the right mix of nitrogen, oxygen, carbon dioxide, and other elements for us and other life to survive. Earth is the only planet in our solar system with an atmosphere that can sustain life as we know it.

The thickness, or rather thinness, of the atmosphere is the result of a balance between Earth's gravity holding it to the planet and energetic molecules that want to rise and move toward space. If the Earth were the size of a basketball, the thickness of the atmosphere would be equivalent to a thin sheet of film wrapped around the ball. This extremely thin, life-protecting blanket of atmosphere is made of layers that extend up from the surface and become less and less dense until the blanket fades into outer space.

The troposphere, the critical, thinnest layer of atmosphere that's closest to the Earth, varies in thickness from only five to ten miles, depending on which part of the Earth you're looking at. It's thinnest in the colder regions, like the North and South Poles. *Tropos* means "change," and in this layer we find a constant mixing of gases that produces changing weather and regulates temperature, our water cycle, and other factors important to life. In fact, all weather phenomena occur in the troposphere, and because nearly all the water vapor and dust in the atmosphere are in this layer, clouds can form. Significantly, almost three-quarters of the atmosphere's density and the most important life-supporting and -protecting features are also contained in this layer within the first ten miles from the surface, which is 78 percent nitrogen, 21 percent oxygen, and a 1 percent mixture of argon, water vapor, and carbon dioxide. And because of its tilt, rotation around the Sun, and distance from the Sun, Earth has diverse regional climates, which range from extreme cold at the poles to tropical heat at the equator. It is this diversity that enables Earth to support a wide variety of living beings. Because solar heat penetrates the troposphere easily, this first layer of our atmosphere also absorbs heat, which is reflected back from the ground in a process called the "greenhouse effect."

The greenhouse effect is a natural process that warms the Earth's surface. When the Sun's energy reaches the Earth's atmosphere, some of it is reflected back to space and the rest is absorbed and re-radiated by greenhouse gases. This process is necessary to sustain life on Earth, but it has also become one of our greatest planetary balancing acts. The atmosphere's most abundant greenhouse gases are carbon dioxide, water vapor, and methane. When maintained in balanced proportions, they support life, but when out of balance they threaten it.

All this life-sustaining activity occurs within the thin blue line. While it is the place where the air we breathe and the weather we experience exists, it is also, very simply, the container in which we live and *why* we can live.

One of my most vivid memories of flying in space has more to do with getting to space than flying in it—the launch. For the ten days before launch, the crew lives quarantined together in a relatively isolated facility called the Astronaut Crew Quarters at the Kennedy Space Center (KSC) in Florida. The primary reason for quarantine is to limit the number of people we come in contact with in order to reduce our risk of being exposed to infectious diseases; if one or more of the crew gets sick, either before or after launch, it could ruin the entire mission. Even a common cold can have a negative impact on the mission, since our immune systems become weakened in microgravity. The problem of congested ears from a cold would be compounded by the sinus congestion experienced in space when body fluids rise upward due to microgravity, and sinus congestion would be particularly painful with the changes in pressure that are required for a spacewalk.

During quarantine, the only people who have any regular contact with us are our medical team, some senior NASA management officials, and the crew quarters staff; we get some very limited contact with our families. Everyone we do come in contact with has been medically screened. Especially difficult for me, though, was the rule that the children in our families have to be fourteen years or older for any contact to happen. (Apparently humans become less germy at fourteen.) My son was too young for contact, at age seven and then nine, before both of my missions. I'm still not at all fond of that rule.

We are meant to spend the majority of our time during the quarantine period in crew quarters. And we do—preparing and studying our checklists, helping with the last-minute logistics for our families and launch guests, and trying to relax a bit by watching movies or exercising. We also take the opportunity to get outside in the fresh air for a daily run or bike ride together. Fresh air is something we know we won't be experiencing during our spaceflight, so time spent outdoors is precious. On one of those bike rides I was pleasantly surprised when one of my very best friends, Jenny Lyons, drove by on her way to work. (Jenny and I had started work together as NASA engineers at KSC a *mere* twenty years earlier.) We followed the rules. I stood on the side of the road with my bike, and she parked her car on the opposite side. We chatted from a distance and gave each other a virtual hug. (This prepared me for the "social distancing" that became a part of all our lives in 2020 during the COVID-19 pandemic.)

Sometime in the last few days before launch, a NASA crew tradition is to hold a "Beach House BBQ" (there is an actual house on the beach at KSC), to which each crew member invites immediate family and four people of their choosing. We are also allowed two or three dinners with our spouses at crew quarters or visits at the beach house, and the day before launch our spouses come with us for a tour of the launch pad and the "wave across the ditch."

Before my two missions, I considered all these opportunities to spend a little time with my husband and family a blessing. I chuckle when I remember these times, though, because I'm sure my husband had very different ideas about how that time should have been spent in contrast to what actually occurred. The reality is that all these planned encounters have an unspoken feeling of "we all know this might be the last time we see you if something goes wrong with launch or while you're in space, but we don't really want to talk about that."

The "wave across the ditch" is one of these encounters that stands out for me. This tradition started back in the Apollo days, when there used to be a ditch around the perimeter of the launch pad where now there is a road. The day before launch, the astronauts would do a traditional wave to their families and close friends gathered across this ditch from them. The "wave across the ditch" was the only time in those ten days before both of my launches when I was able to "see" my son. I have to admit, I "went a bit rebel" both times and scurried across the street, locked eyes with my son, and quickly grabbed his hand. Both times he gave me the sweetest look, and I secretly passed him a small seashell with MLR (for Mommy Loves Roman) written inside.

Our time in quarantine also helped us further prepare as a crew in those final days before the mission by providing a comfortable and quiet living environment, great food, exercise equipment and personal trainers, and the proper lighting to help us get our circadian rhythm in sync with the time of day for launch and the mission. My first launch was originally scheduled at 4:00 a.m., but after several scrubs we ended up launching one minute before midnight; my second launch was at 4:53 p.m. Nothing about being an astronaut fits into a normal nine-to-five schedule.

We always arrive at the launch pad several hours ahead of the scheduled launch. Before my second mission, at one of our stops on the van ride to the pad, I poked my head out the door and waved back at the people wishing us well along the way. I saw so many familiar faces, including the people I'd worked with at the Kennedy Space Center for ten years to help prepare everything for other astronauts before they took off into space. Now, on this day, they were waving and cheering for me. It's one of my favorite memories.

When we finally arrived at the pad and stepped out of the van, we all looked up in awe at the sky and wondered what it would be like to blast through it on our way to space together, and we also looked up at our spaceship towering a couple hundred feet above us. It seemed alive, with steam pulsing from the active systems that were warming up and getting ready for the final countdown to launch. We rode the elevator together up 195 feet to the level with access to the Space Shuttle hatch. Before approaching the hatch, we took turns taking one last potty break in the designated launch pad toilet (another launch-day tradition).

As you might imagine, those orange spacesuits weren't the easiest to maneuver in a tiny bathroom closet, but I was grateful for the opportunity for that relief before strapping in. We also had a minute to make a quick phone call to our families before walking across the metal-caged access arm to the hatch of the spaceship. Then we entered an area called the white room and met up with the team known as the closeout crew. They helped us finish putting on all the other pieces of our spacesuit and smacked each of us on the back to wish us well. Because there were cameras monitoring all this, we were able to pass silent messages to our families. My son and I had come up with a secret series of hand signals. Finally, one by one, we climbed in and were helped into our seats. Lying on our backs, we were strapped in snugly.

Once I had put my checklists and procedures in their proper place and we all had done some communication checks with the launch control team, we had some time to relax, so I took a nap.

I didn't really begin to believe I was actually going into outer space until about twenty minutes before launch, when the crew became more actively involved with the countdown. But when we got to the iconic "ten, nine, eight..." point, that's when I thought, *Here we go!*

When the countdown gets to six seconds on the Space Shuttle, the fuel starts flowing from the big orange external tank to the three main engines on the back of the orbiter. Those engines together consume liquid fuel at a rate that would drain an average family-sized swimming pool in under twenty-five seconds. The Space Shuttle is made up of four different parts: the orbiter (the spaceship part), the orange external tank, and the two white solid-fuel rocket boosters strapped to the side of the tank. When the orbiter's three main engines lit, we heard a rumbling sound, and because the engines were at an angle on the back of the orbiter and we were sitting on top, we felt the whole stack of the Space Shuttle rotate forward about six feet. The stack bounced back into place and all went vertical again when the countdown struck zero. NASA uses a highly technical term to describe this whole motion: it's called "the twang." With perfect synchronization, as soon as the stack was vertical, the solid rocket boosters lit, and I felt like we'd been kicked off the launch pad.

My mind flashed back to all the years of preparation that had gone into getting to the point of strapping into that seat for launch. Then I flashed to the thought of my family watching and what it must be like for them. I was amazed at how quickly we were lifting off the pad and well on our way to space. All I really knew at that moment was that I was moving really, really fast. In just eight and a half minutes, we went from stillness on the launch pad to traveling at 17,500 miles per hour in order to achieve orbit around our planet in space (nine times as fast as the average bullet). I'm still surprised by how much a body can shake and still remain intact during those eight and a half minutes.

My first launch was at night, and I was on the mid-deck (the lower level of the shuttle cockpit), so I had no windows to look out of. But my second launch happened during the daytime, and I was on the flight deck, where I could see out the windows. The blue of the sky in the light of day shone through those windows—until it didn't. Nothing prepared me for the physical and emotional dynamics of launching to space, and nothing prepared me for how quickly the view out the window goes from blue to black. Seeing that instantaneous shift helped me understand just how thin that blue line of our atmosphere really is.

I marveled at not only the beauty but the strength of that remarkably thin blue line that protects us. I was also in awe of the strength of the Space Shuttle as it withstood the immense forces required to push out of Earth's gravity, well beyond the density of our protective blue atmosphere, and make it safely through to the black void. There our survival and protection became solely dependent on the spaceship's thin metal hull and the environment created for us inside it.

I would love it if all of you who want to travel to space could do so and experience the view out the window. You'd find that one of the many wonders of circling the Earth every ninety minutes is that each time you get halfway around, every forty-five minutes or so, you are presented with the sight of a beautiful sunrise or sunset. If you're lucky enough to catch the terminator line—where night is just about to turn to day, or day to night—you can see the edge of the shadow between dark and light flash across the planet; at the same time, you can watch the light that outlines the horizon transition through all the colors of the rainbow before the Sun rises into view or sets into blackness.

Whether witnessing the transition through sunrise or sunset, the one thing you would always see first is the glow of Earth's thin blue line of atmosphere. I was startled when I saw the contrast between the width of the planet and the thinness of the atmosphere for the first time. The atmosphere looks as if it's just a thin blue veil wrapped around the planet.

Seeing this from space, I immediately realized the similarity between the thin blue line of Earth's atmosphere and the thin metal hull of a spaceship. Whether on Earth or on the ISS, these protections are the only thing between us and the deadly vacuum of space.

We built a life support system in space to enable people to live and work there. Nothing—no science, no maintenance, no routine daily operations—can be performed if the crew cannot survive; beyond that, the systems must also be robust enough for the crew to thrive. Like the space station, the Earth is a closed-loop life support system. If it is to continue to do its job, we must operate within the bounds of the resources it contains, and we must adequately maintain those resources, in both quality and quantity, in order to survive.

Earth's thin blue line of atmosphere is a perfect analog to the thin metal hull of our spaceship, and the way we preserve our environment on the ISS is a simple example of how we should be doing the same on our planet. Here on Earth we tend to take our natural life support systems for granted, forgetting that our planet is like a spaceship. To survive space travel, however, we have to mimic each of these systems that supply clean water, clean air, comfortable temperature and air pressure, and so on, by creating comparable mechanical systems.

Thus, on the ISS, we are especially aware of how critical the thin metal hull is to our survival. Every moment of every day, we are mindful that we need to manage all the resources contained within it, as well as the integrity of the hull itself, in order to survive. Every day we monitor the amount of carbon dioxide and oxygen in our self-contained atmosphere, and we remain on the lookout for anything toxic in our station's precious air supply; all the while, our mission control team on the ground is helping us monitor these resources, as well as any known debris in space that might hit us (and we are aware that something unknown might hit us too).

The simplest way to describe what the ISS metal hull and the Earth's atmosphere have in common is that they hold in all the good stuff. The downside is that they both do a pretty good job holding in the bad stuff too, and if the stuff inside gets bad enough, the life they are intended to support cannot survive, and the thin protective layer can be damaged.

On the space station, the metal hull holds in the air and keeps it at a level of pressure that allows us to breathe. A crack in the hull, allowing the air to escape to the vacuum of space, would do critical harm to us, the crew, though the station itself would not necessarily be seriously damaged. With no air pressure, not only would we suffocate, but our bodies would start expanding, and the water that makes up most of our bodies would start to vaporize. Any moisture in our mouths would boil away on our tongues. Our bodies would eventually freeze, but we wouldn't know it because we'd already be dead. Not a good day in space.

On Earth, would you ever consider a scented candle or a paint chip a threat to your existence? On the ISS, these are very real threats. We do our best to reduce the risk of something happening that would trigger the emergency alarms by proactively managing our life support system from both the inside and the outside. Life on the ISS depends on the delicate balance of the environment that must be maintained at all times against internal threats to the quality of our air and external threats to the integrity of the hull of our spaceship. These threats are magnified by the unavailability of any air except the air created by machines that are part of the ISS itself; outside its thin metal hull lies only the deadly vacuum of space.

We bring toxicity into our homes here on Earth all the time through items treated with chemicals that can cause us harm. We light scented candles, install new carpet, paint our walls, and spray deodorizers, which cause a phenomenon known as off-gassing. Off-gassing is the release of airborne particulates or chemicals known as volatile organic compounds (VOCs), but a simple definition makes the most sense to me: off-gassing is "when a material releases a potentially harmful chemical [VOC] into the air."

Even though the acronym VOC contains the word "organic," that doesn't mean VOCs are good for us; it just means that these compounds contain carbon. The greater problem is related to the word "volatile," which is a way of describing how easily a substance evaporates at normal temperatures. VOCs range from harmless to mildly irritating to cancer causing; they also range from odorless to a strong odor, but the level of its odor is not necessarily an indication of a VOC's toxicity. Individual effects will vary depending on the type of VOC, the concentration in the air, and the length of exposure time.

On the space station, off-gassing is one of the main reasons why many kinds of materials are either banned or restricted to minimal use. To manage air quality, we must limit off-gassing to prevent the occurrence on our space station of what's been referred to as "sick building syndrome." Most everything on the space station is unscented. The only real smells you encounter are from yourself, your crewmates, the food, the trash... and the "smell of space."

Of course, for obvious life-threatening reasons, you can't actually "smell" space, but there is a distinct odor to materials once they are brought into the space station after exposure to the vacuum of space. Just like seeing those shooting stars below me, I was never told before flying to space to expect this unusual smell. I always noticed this distinct smell when a crew member came back into the airlock after a spacewalk, or when we got first access to the hatch of a recently docked spacecraft. Not unpleasant, the smell reminded me a bit of the sweet, metallic odor of an overheated car radiator.

It turns out that this odor known as the "smell of space" is not the smell of space at all, but the result of off-gassing caused by oxidation. In space, atomic oxygen (single atoms) can cling to materials like a spacesuit's fabric, to tools, and to the hatch of a docking spacecraft. When these single atoms of oxygen combine with the atmospheric oxygen (O_2) in the cabin, they make ozone (O_3), which is what we're told causes the odor.

Ever since the earliest spaceflights, NASA has been figuring out ways to protect astronauts from materials that off-gas. Unlike buildings on Earth, spaceships have no way to let in fresh air. Like all spaceships, the space station is purposely built to be airtight, so airing things out by opening the window isn't an option on the space station like it is in our earthly homes.

Any materials used to build the interior of the space station, all the components used for science experiments, and anything that is sent to space for the crew (including our clothes and hygiene products and even my watercolor kit) must be evaluated first, not only for flammability risk, but also for its potential for off-gassing and toxicity to the crew.

In addition to managing the threat of a toxic atmosphere inside the space station by proactively restricting the types of materials that are allowed into it, the space station's air system is also continuously filtering and "scrubbing" the air to keep it clean while also monitoring the constituents in the air (including the necessary components of breathable air like oxygen and nitrogen, but also harmful chemicals like benzene, methanol, and toluene) to ensure that all are at safe levels for the crew. If any of these constituents are out of healthy ranges, an alarm sounds.

These same concerns exist here in our homes on Earth, and the technology developed for the space-based air quality monitors and filtering system has now been adapted for home-based systems. These systems, which are just starting to become popular, will help make it easier for us to understand and manage the air quality in our homes and workplaces. Even though we can usually open a window to get some fresh air flowing, the presence of VOCs is no less threatening to us here on Earth than it is in space. And just as we have done with our spaceships, we can make choices about the kinds of materials we bring into our homes.

Another interesting spin-off from space-based technology developments that improves indoor air quality down here on Earth came indirectly from the study of crop growth on the space station. The plants are being grown as a test bed for what we'll need to do once we journey to and live on Mars. Scientists working on this investigation noticed that a buildup of a naturally occurring plant hormone called ethylene was destroying plants within the confined space station plant growth chambers. As a solution, they developed and successfully tested an ethylene removal system in space that also removed viruses, bacteria, and mold from the plant growth chamber. Commercial companies then adapted the system to purify air here on Earth. The technology is being used around the world in hospitals and other medical facilities to improve air quality for patients by removing harmful germs and bacteria, in food trucks and grocery stores to prolong shelf life, and in wine cellars to reduce mold; it is also being used to clear the air of mold, mildew, germs, and unwanted odors in everyday living environments like our homes.[1]

The main external threat to the space station's thin metal hull, which holds in the entirety of the life-sustaining environment, is micro-meteoroids and orbital debris.[2] NASA loves acronyms, even for the stuff that might hit our space station, so we call it MMOD.

MMOD (micro-meteoroid orbital debris) consists of a combination of millions of naturally occurring micrometeoroids and man-made debris in space. On the one hand, micro-meteoroids seem pretty cool because they are of extraterrestrial origin, which means they come from outer space, where they travel in an orbit around the Sun. If a micro-meteoroid's orbit intersects with Earth's and it makes it through our atmosphere to the ground without burning up, it's called a micro-meteorite. On the other hand, micro-meteoroids are not so cool because, like the shooting star I saw from the window on the ISS, there's always the risk that they'll hit our spaceship.

Further, there's a growing threat from the man-made debris in space, commonly referred to as "space junk." Space junk ranges from big stuff like parts of rockets, spaceships, and satellites to very small debris like flecks of paint from these spacecraft. All this junk that has yet to reenter and burn up in Earth's atmosphere is orbiting the Earth out there with us and our spaceship. If orbits align, we might get hit. Traveling at 17,500 miles per hour, even a fleck of paint can do damage.

Here on Earth, several scientific and military organizations detect, track, and catalog the debris in space. The primary responsibility for this in the United States falls to the US Space Surveillance Network, which is part of the US Space Force. While doing this work of detecting, tracking, and cataloging, these organizations also monitor anything that might be a threat to our spaceship. They track the more than 23,000 known man-made fragments larger than about four inches as they orbit around our planet. Those are the pieces large enough to track; it's estimated that there are more than 500,000 pieces between 0.4 inches and 4 inches across that, to date, are not trackable. Thankfully, though, objects as small as 0.12 inches can be identified by ground-based radars, allowing scientists to at least statistically estimate the population of the smaller pieces, even though they can't track them. Most of this debris circles within 1,250 miles of Earth's surface in what is known as low Earth orbit (LEO), an area that is home to lots of satellites, including the International Space Station. (The ISS orbits somewhere between 200 and 300 miles up.)

Given that the Earth is surrounded by this orbiting debris, getting a rocket launched through it and safely into space requires careful consideration of when and where to launch. For NASA launches, a group of engineers known as flight dynamics officers are responsible for creating a "launch window" that sets the safe timing for rocket launches. They calculate a window of time based on two primary criteria: where the rocket needs to arrive in space (to dock with a space station, for example, or to land a rover on Mars) and what trackable debris or other spacecraft might be in the way.

In addition to tracking the MMOD, we also have special shielding on the ISS to help protect the metal hull, and based on tracking input from the ground, we can boost the station up or down in orbital height to avoid potential impact. In the worst case, we can take shelter in our Soyuz spacecraft (the Russian capsule that is also our rescue vehicle) and return to Earth if the threat of being hit becomes highly likely. (NASA and our international partner agencies have agreed upon criteria for what is considered "highly likely.")

There is no denying that these threats are very real and that we need to actively manage them. Nevertheless, all these measures and more are in place to manage threats to the crew's survival, whether from something toxic in the air or a hole in the hull, with the best possible information and technology.

Consider this analogy with respect to Earth's thin blue line of atmosphere. Like the hull of the ISS, our atmosphere holds the oxygen and other good stuff in and keeps dangerous radiation out. If the thin blue line were to be severely compromised, Earth would still exist, but our planet would no longer support life—any life.

In the upper atmosphere, there is a transparent and protective layer of ozone. Ozone (O_3) is a highly reactive gas composed of three oxygen atoms. It is both a man-made product and a naturally occurring molecule found in the Earth's stratosphere (the upper atmosphere) and troposphere (the lower atmosphere). *Where* ozone is in the atmosphere means very different things for life on Earth. Ozone in the upper atmosphere is vital to our survival and to the survival of all other species living on the sunlit Earth while ozone in the lower atmosphere is toxic and harmful.

Ozone in the stratosphere acts like a natural layer of sunscreen for us and the planet. Without this protective layer in our upper atmosphere, deadly levels of cancer-causing ultraviolet radiation from the Sun would reach us on the ground. In the mid-1980s, this transparent layer of protection, which scientists had been monitoring over Antarctica, became alarmingly thin. Shortly afterward, through colorful NASA satellite imagery, we were able to obtain a picture that clearly showed what became known as the "ozone hole"—a breach in the thin blue line of the "atmospheric hull" of our planetary spaceship.

The science leading to the discovery of this damage to the ozone layer resulted in the Montreal Protocol, which, as of this writing, is the most successful example of international cooperation to protect life on this planet so far. It has been recognized with the highest scientific honors, including the Nobel Prize in Chemistry, which was awarded in 1995 to Drs. Paul Crutzen, Mario Molina, and Sherwood Rowland for their work in atmospheric chemistry, particularly concerning the formation and decomposition of ozone.

Their work provided direct evidence of the damaging impact of human activity on the integrity of our atmosphere: they conclusively showed that the use of chlorofluorocarbons (CFCs) (the freon and aerosols with a wide variety of different applications in everyday life, like refrigeration) depletes the life-protecting ozone in the upper atmosphere, and that pollutants generated from motor vehicles and other combustion systems create toxic ozone in the lower atmosphere.

Dr. Mario Molina, a chemist from Mexico and one of the Nobel Prize winners that evening in 1995, passed away in October 2020 after a lifetime of scientific discoveries and work to apply in support of solutions to our planetary challenges. In 2004, he opened the Mario Molina Center for Strategic Studies of Energy and the Environment in his hometown of Mexico City (and in 2005 at the University of

California at San Diego) to serve as a bridge between science, public policy, and business sectors for resolving environmental and energy problems.

In his later years, the climate emergency became an increasing area of focus for Dr. Molina, along with the associated issues of air quality and air pollution. Molina continued his Nobel Prize–winning research with attention to some of the CFC alternatives that are also powerful greenhouse gases: some, such as hydrofluorocarbons (HFCs) and halon, could warm the planet at a rate many thousands of times greater than carbon dioxide could.[3] He was a force behind a number of amendments to the Montreal Protocol that have phased out these alternative chemicals and brought even greater protection of our planetary life support system.

In Dr. Molina's own words, "we had decided to communicate the CFC/ozone issue not only to other scientists, but also to policy makers and to the news media; we realized this was the only way to ensure that society would take some measures to alleviate the problem. I am heartened and humbled that I was able to do something that not only contributed to our understanding of atmospheric chemistry, but also had a profound impact on the global environment."[4]

In response to the news of Dr. Molina's death, Durwood Zaelke, president of the US-based Institute for Governance and Sustainable Development, who worked with Molina in urging governments to take action on the climate, said: "The Montreal protocol solved the first great threat to the global atmosphere, and has done more to solve the next threat—the climate threat—than any other agreement, including the Paris agreement. Phasing down CFCs and related fluorinated gases has avoided more climate warming than carbon dioxide is causing today. Mario remained deeply involved until his last days."[5]

It took courage for Molina, a chemist from Mexico, and Rowland, chair of the Chemistry Department at the University of California at Irvine, to speak out about their findings. The industries that made and used CFCs put up a strong fight in response, and even some in the scientific community kept their distance for fear of controversy.

Since the discovery of CFCs in the late 1920s, they had been used extensively across the globe in consumer goods like aerosol cans and as refrigerants. Being inflammable and nontoxic to human beings also made CFCs very popular substances. But as Molina and Rowland discovered, CFCs rise miles up into the Earth's atmosphere after they are released, and once in the stratosphere they are split apart by sunlight, forming chlorine atoms that destroy the protective layer of ozone molecules. (Just one atom of chlorine can destroy 100,000 ozone molecules.)

It took thirteen years of advocacy by Molina and Rowland and many other supporters in the scientific community, at nongovernmental organizations, and within the media and the public before the world responded. Adopted on September 15, 1987, the Montreal Protocol, an international treaty put in place to ban man-made CFCs, is to date the only UN treaty ever to be ratified by every country on Earth—all 197 UN member states. As a result, 97 percent of all ozone-destroying chemicals have been eliminated worldwide. There has been much improvement, and while the ozone hole still opens over Antarctica each September, it is expected to disappear sometime around 2050.

Former UN secretary-general Kofi Annan described the Montreal Protocol as "the single most successful international agreement to date."[6] It is an impressive example of international cooperation in response to a global threat; unlike other recent international agreements, it not only binds countries to action but contains financial provisions to assist them in doing so. Like the scientific platform and international cooperation we see with the ISS, the Montreal Protocol provides proof that meaningful cooperation is possible on a planetary scale. In the United States, President Ronald Reagan played a leading role in advocating for the Montreal Protocol and a ban on CFCs, and he also played the key role in directing NASA to build the International Space Station.[7]

The results of the CFC ban have been both promising and sometimes puzzling. Promising because the Montreal Protocol has unquestionably been a cooperative success, resulting in significant progress in the wide-scale reduction in the use of CFCs toward the goal of completely eliminating their use by later this century. But puzzling because scientists are finding that, even as there has been measured improvement in the ozone layer, the overall composition of the atmosphere is still at risk.

By learning about the effects of man-made chemicals like CFCs and their interaction with our atmosphere, we are in a better position to understand how other chemicals and human behaviors interact with our environment as a whole. We live in the troposphere, Earth's lower atmosphere, and it is the place where ozone goes from being a protective barrier (as in the upper atmosphere) to becoming a threat. Ozone in the troposphere is a harmful air pollutant. One of the six common air pollutants identified in the Clean Air Act, it is harmful to all types of cells. Ozone is the main ingredient in smog. When it's inhaled, ozone can damage lung tissues, which is one reason why most of the developed world monitors (and reports) on air quality in real time every day.[8] Ozone is produced when pollutants emitted by cars, power plants, industrial boilers, refineries, chemical plants, and other sources react in the presence of sunlight. These same pollutants are primarily responsible for climate change.

Ozone depletion affects climate, and climate change affects ozone. The successful implementation of the Montreal Protocol has had a marked effect on climate change. This is cause for hope. The legacy of the CFC/ozone depletion story and the Montreal Protocol is proof that effective environmental action *can* be taken at the international level to resolve our greatest environmental problems.

Air, sea, land, and space are all interconnected, as is all the life that inhabits these places—all life is protected and contained by the thin blue line. In the same way that a breach by CFCs in the atmosphere over one part of the planet has global effects, anything that has a negative impact on the integrity and health of any place on our planet negatively affects the whole. Just because we might not "see" that effect doesn't mean it isn't happening.

While the interconnectivity of air, sea, land, and space was obvious to me from the vantage point of a space station, I am pleased when I discover this same sense of interconnectivity through experiences here on Earth.

In 2019, my family and I took one of our regular trips to the Isle of Man, a small, beautiful island with rolling green hills and medieval castles that sits in the middle of the Irish Sea between England and Ireland. If you know the island at all, it's probably for its TT Motorcycle race, tailless cats, or reputation as a leading global finance center. I had no clue the island even existed until I met my husband in the mid-1990s, and since then it has been one of my favorite places on the planet, and the place my in-laws (two of my favorite people on the planet) still call home today.

Beyond its stunning beauty, the Isle of Man has a long history of being a progressive and resilient place. In 2020, its inhabitants celebrated 1,041 years of continuous parliamentary government under the High Court of Tynwald, its legislative body. The island's motto, "Quocunque Jeceris Stabit" (Whichever way you throw me, I will stand), dates back to AD 979, in Viking times. It was the first place in the world to grant women the right to vote—thirty-seven years before the United Kingdom, and thirty-two years before the United States.

During this trip to the island, I met Rowan Henthorn, a marine scientist and ocean advocate who had lived on the island all of her twenty-seven years and was working for the Isle of Man government as a climate change researcher and ecosystem officer. In 2018, she was a member of the crew of the first eXXpedition North Pacific mission to what is known as the Great Pacific Garbage Patch.[9] The aim of the on-going, all-female eXXpedition sailing voyages, which I had followed from afar through social media, is to "raise awareness of, and solutions for, the devastating environmental and health impacts of single-use plastic and toxins on the world's oceans. To make the unseen seen."

Like most people, when I first heard of this place called the Great Pacific Garbage Patch, I imagined a massive floating island of garbage, perhaps even the size of the Isle of Man, on the surface of the ocean. My imagination was not imaginative enough because, while not an actual island, the garbage patch is estimated to be twice the size of Texas—about 2,500 times larger than the Isle of Man! Regardless, I wondered if I could have seen it from space (I could not). In reality, these patches are largely unseen. Flying just above it or sailing through it, you do not see an island of garbage at all, only some cloudiness in the water and smaller clumps of floating trash. This is why it's more appropriately described as an oceanic smog than an island. These garbage patches are out of sight in the middle of the Pacific Ocean, so they often remain out of mind, but that doesn't make them any less problematic.

Figuring out how much plastic is really there and how to clean it up is made even more complicated by the fact that the bulk of the debris is teeny tiny—what's known as "microplastics." *Trillions* of pieces of microplastics are mixed with larger pieces of debris, such as discarded fishing equipment and nets and other larger varieties of trash. All this moves through the ocean in the swirling currents known as a "gyre." So, this garbage is not only an issue on the ocean's surface but is distributed throughout the water column. Microplastics have been discovered polluting the deepest explored parts of the ocean (and at the tops of the highest mountains).

I found it difficult to wrap my mind around this amount of microplastics infiltrating our oceans. I've seen the models that scientists have built to represent how many satellites and how much space junk is circling our planet, which is difficult enough for me to comprehend, but that pales in comparison to the millions of times more plastic in the ocean. To me, it's overwhelming.

Rowan and I spoke of her commitment to meaningfully share her eXXpeditionary experience of sailing in the midst of this plastic and to bring it to life for her fellow islanders at home. She is working hard at both the governmental and grassroots levels to draft the island's policies for plastics and its response to climate change, and to raise awareness and encourage sustainable improvements to the quality of life both on and off the island. She also helped me become aware of the surprising number of positive environmental protection and sustainability initiatives happening on the Isle of Man.

I asked Rowan if she thought that people would be more encouraged to take action if the garbage patch really was a massive island of plastic floating on the surface of the ocean.

"They already think it is, so I'm not sure there would be a greater response, unless maybe if it was floating in their own backyard." Then, with a little more excitement in her voice, she added, "When you are in the middle of it, you look out on a blue expanse of water around you, for as far as the eye can see it's blue. Yes, there are many recognizable plastic items floating past the boat, but in large it's blue. It's not until you put a net in the water that you realize that you are surrounded by trillions of tiny, microscopic pieces of plastic.

"For me that's much scarier than a plastic island. A plastic island in a way seems manageable—scary, but manageable. That's why it's so important to make the unseen seen. I hope it helps people to realize that the solutions to issues like this start on land, with our actions. There is no silver bullet that's going to suddenly scoop all the plastic out of the ocean. We have to change our systems and our behavior—we have to turn off the tap!"

You'd think that traveling to space or sailing three thousand miles across the ocean would allow you to truly "get away," but it doesn't.

As Rowan put it, "There is no 'away.'"

There's no "away" for us because no matter where we go, even to these remote places, we still experience the impact of our own human behavior, and the evidence of our behavior is presented to us through the stuff that doesn't go away, like the plastic in our oceans. There is continuity in all this. I think it all comes down to the interconnectivity of everything that surrounds us: we live on a planet surrounded by unseen MMOD, protected by the thin blue line of atmosphere that surrounds and contains the air, land, and sea, and all the life that inhabits these places, even the microplastics permeating our oceans. All is interconnected, and all influences the planet's ability to sustain life in one way or another.

Just as seeing Earth from space gave me a new perspective on who and where we all are—together on this shared planet—leaving the Isle of Man for the far reaches of the Pacific Ocean gave Rowan a similar new perspective. This delighted me for many reasons, including the confirmation that you don't have to leave Earth on a spaceship to appreciate our unity as Earthlings on this planet. We spoke of our surprise at the ability of these opportunities to experience extraordinary places to broaden our perspective and bring us a greater sense of home, to remind us that we live together on one planet.

I asked her if being on a sailing ship in the middle of the ocean made the fact that we live on a planet any more real to her.

"Yeah," she said. "I think especially lying in my bunk at night because that was really my time to reflect on where I was. The stars were so incredible, and being surrounded by the cosmos really made me think about the planet we were sailing across. I often would lie in my bunk and kind of zoom out in my mind's eye and go further and further up to imagine just this tiny boat of girls traveling across the ocean. It felt big in the sense that we were traveling across the ocean and continually moving at speed, twenty-four hours a day. My world also felt quite small in the sense of realizing you could even be in the middle of the ocean and that your 'world' is also complete in just the five meters of ocean that surround you. And actually living in such a connected world in general, makes it feel quite small, makes it feel like you have the potential to really connect to people all over the world, which is exciting!"

In a comment that reminded me of the ISS—where the relationships we have developed with our partner countries have brought such great success, and where I gained so much from the relationships I established with my crewmates—Rowan said that the eXXpedition trip "was the most incredible adventure. But actually one of the really (if not the most) special things about it was meeting women from all around the world and building a connection, exploring an issue, thinking and talking about solutions together, and figuring out where you as an individual fit into that puzzle. That was such a real and beautiful thing to come out of it."

When I asked if she'd consider anything about her experience an Earthrise moment, she replied, "I like to think that as you go through life you have many Earthrise moments because your world is always changing. Probably my most recent Earthrise moment was realizing that it's okay to back yourself a little bit more, and the realization that probably the only person who thinks you can't achieve something is yourself. Spending time with all those incredible women helped me to realize that."

Expanding on what she learned about herself, Rowan added, "I returned to the island with this renewed sense of what I can do to protect this planet we all live on. Realizing that a lot of people are on that same journey, and really, we are all doing the best we can with what we have at any given moment. I think figuring that out helped me to treat myself and others with more compassion, which is really important, especially when it comes to creating positive environmental change."

When astronauts leave Earth for space, we all understand that we are heading off the planet in support of the space program's mission. We believe in this mission because it is based on a pursuit of knowledge for the greater good. It acknowledges our human need to challenge ourselves and to explore, and we do so for the benefit of life on Earth. We all also know, even before we go, that this will be a life-changing experience. As a result, we each discover and come back to Earth with our own personal mission and a need to share the experience, to take action in response to it, and to encourage action from others. Going to sea with her crew on the eXXpedition mission did the same for Rowan, and I think this might be why I feel such an affinity to her.

No matter where we are, in space, at sea, or at home, most of us wonder how anything we do on our own could have a meaningful impact on the massive problems that surround us.

"Yeah," Rowan retorted. "'*What difference can I make?* said seven billion people.'"[10]

"Wow!" I chuckled, and I thought about how the truth in these words is so simple, yet so obvious and so powerful.

She went on to say, "I think we'd solve a lot of the world's problems if we all realized that everybody's got something really important to give."

I agreed that we all need to understand that we each have the power to help make life better. We discussed this some more, and Rowan noted that our doubt in our own abilities as individuals can also be felt by an island population and may be why we feel so overwhelmed when we think about overcoming challenges for a planet as well.

"I think one of the risks when you live on a small island is the belief that you're too small to make a difference," Rowan said. "But actually, being a small island means having the opportunity to change and adapt quickly, so it provides a great test bed for new environmentally friendly ideas and initiatives, which can potentially help to find solutions to the global challenges we face."

Rowan told me that one of her greatest challenges after coming back to the island was figuring out how to share her incredible eXXpedition experience in a way that would connect with her fellow islanders in order to drive meaningful change. This sounded so familiar—I've felt the same after coming back from space.

"Spending all that time at sea made me realize that the solutions to these issues really start on land. So when I got back, I was really keen to come and try to create some positive environmental change back on the island that first inspired me. I was really lucky that the Isle of Man government offered me a position in their environment team, working to reduce plastic across all government departments and the wider community."

One of the ways Rowan has worked to reach out to as many people on the island as possible is through community talks. When she discovered, however, that the passion she felt for the subject wasn't necessarily translating to her audiences, she thought that maybe she was focusing too much on the science of the problem and not enough on the emotion she'd experienced during the eXXpedition. Searching for some help with this, she went to a retreat where she was able to explore the human response to the climate crisis from a place of thought, feeling, and connection.

I felt a similar challenge after coming home from space and recall how helpful it was to discover that the story we can share about our own experiences is how we best connect to others. People want to know things about my spaceflight experience that help them establish their own relationship with it; only then does the science of it become important to them.

"Now, at the end of my talks, I just try to keep it real," Rowan said. "I fully acknowledge that these are such huge, scary issues and topics [climate change, single-use plastics], and I encourage the audience to sit with that feeling for a minute.

"I tell people that's okay, that it's good to respond with emotion, to feel it. I don't try and make it any smaller or give solutions that feel insufficient, but I try to encourage people to use whatever they are feeling, often fear, to light that fire in their belly. To help galvanize action and create the change needed to move forward in a more sustainable way. To really feel it; to feel it in your stomach and your heart, and accept that it might make you feel a bit sick or worried, or angry or sad. You just have to try and not let that paralyze you into inaction, but really be the fire that stokes you into action, and then you'll seek out the solutions that are relevant for you."

In 2017, Rowan founded Sustain.Our.Seas (S.O.S), a nonprofit that seeks to reintroduce people at a community level to the nature around them. She believes that "the key to creating change and raising awareness of the environmental challenges we face is to help people reconnect with nature. With Sustain.Our.Seas, we tend to focus on more creative projects like art exhibitions and immersive ocean experiences. I think once you instill a connection with the natural world, people care enough to think about the environmental challenges it faces. It becomes more personal, and with that comes the ability to create positive change in their own lives."

Her government work also has been about connecting the government with the community at large. "We have sustainability champions who drive positive change in each government department, and by 2021 we hope to have removed all unnecessary single-use plastic across government."

"I'll be checking in to see how this is going," I said.

Rowan welcomed that, then went on to describe how the government has developed a community plastics plan that focuses on trying to reduce single-use plastics across the island, in businesses, at home, and at events.

"As part of the plan, we recently set up a 'Plastic Free Isle of Man Working Group' with many of the amazing NGOs and businesses that are passionate about the topic and have a wealth of knowledge. We want to work together to share resources and generally have a more cohesive approach to tackling single-use plastic use across the island."

Rowan provided an example: "The National Sport Centre has fully engaged with the Government's Plastics Plan, having removed plastic bags for wet gear, and plastic overshoes for the changing rooms." Rowan estimated that this change would be similar to "removing approximately ten-thousand-plus plastic items per year." And most major island events, like the TT Races, the Tynwald Day Fair, and the Food and Drink Festival, have banned the use of single-use plastic items for vendors. Water refill stations around the island, including in schools, have resulted in increased use of reusable alternatives to plastic bottles and has people considering alternatives to single-use plastics in general.

The people of the Isle of Man have been tackling challenging problems since the Viking days. All of the islanders I know truly appreciate the gift of natural beauty and bounty they are blessed with. This is true for the many, like Rowan and my in-laws, who were born on the island and choose to stay there, as well as for those who have come from somewhere else to make the island their home and those, like my husband, who have moved off-island but consider the good of the island in everything they do. They have been working hard to find sustainable solutions for the island's people, environment, and economy, and at the same time also for the planet.

I have become enamored by the beauty of the island and value so many things about it, like its ranking as Europe's number-one location for dark sky astronomy. Within the island's mere 220 square miles, there are twenty-six designated dark sky sites. On a clear night, I'd look up and wonder, *How can there be so many stars in the sky, but I can't see my hand in front of my face?*

While the island is a place of extreme natural beauty, it also has a reputation as a world-class financial center, and I was pleasantly surprised, shocked even, to discover the world-renowned technical expertise that exists within this small island community.

The Isle of Man is home to Ronaldsway Aircraft, the company that builds ejection seats for high-performance military aircraft for ninety-three air forces around the world (including the T38 trainer jets I flew as an astronaut). It is also home to Manx Precision Optics, which handcrafts the highest-quality optics for spacecraft (including the rovers on Mars), and to Swagelok, which manufactures and supplies critical fittings for the ISS and before that for both the Space Shuttle and Apollo programs. The Isle of Man was the test bed for the first 3G communications in the world and is a leader in the business of satellite communications and commercial spaceflight (most notably with Isle of Man satellites providing broadband to over four billion people around the world).

Like the ISS, which is a shining technological masterpiece that's largely unknown and unseen as it flies through space with its international crew, peacefully and successfully performing work together to improve life on Earth, the Isle of Man is a small island gem that is generally unknown and unseen in the middle of the Irish Sea, a community whose people are peacefully and successfully working to sustainably conserve and grow their environment and economy in harmony with their place on the planet.

All of this is to say that it came as no surprise to me when I learned that in 2016 the Isle of Man had been designated as the only *entire-nation* UNESCO Biosphere. This was the first time in the fifty-year history of the UNESCO Biosphere program that an entire nation has been honored with this designation. At the time of this writing, the Isle of Man is still the only entire nation to be so designated (in all, over 700 sites in more than 125 countries are Biosphere Reserves).

UNESCO (United Nations Educational, Scientific, and Cultural Organization) established the Biosphere program in 1970 under its Man and the Biosphere (MAB) program. The Biosphere program initially concentrated on experimenting to find solutions to care for the land, the sea, and the species. It was expanded in the mid-1990s to also care for the designated areas of economy, culture, heritage, and community, which is right up the Isle of Man's alley, especially given that even before the Biosphere designation the island had already pledged to conserve its amazing landscape, wildlife, culture, heritage, and communities and to develop its infrastructure and economy in a way that would support the environment and raise local awareness of what made the island so special.

The Manx, as the people of this small island nation (population eighty-five thousand) are called, have a strong cultural connection with both the sea and the land via fishing and farming, so they naturally are aware of their reliance on one another and their natural resources. They understand the limitations of their own resources and their interdependence with the world around them. An independent nation, they have established peaceful and successful relationships with Great Britain (they are a crown dependency and part of the British Isles) and other countries around the world.

There is no UN requirement to do so, but the Isle of Man has chosen to leave management of its Biosphere activities to its Department of Environment, Food, and Agriculture (DEFA) because the Manx believe that rapid progress can be achieved through this integration of government and community. Similarly, everything from children's educational and community activities to policies established by the financial and technical companies on the island are leveraged to implement new strategies and policies based on the Biosphere philosophy, which is to promote solutions that establish a synergy between the conservation of biodiversity and its sustainable use.

The island's scouting organization worked with UNESCO to create the first Biosphere Scout Badge, for which the scouts must complete tasks relating to the UN's seventeen global Sustainable Development Goals (SDGs) for a better world. Thus, scouts on the island have initiated or become involved with planting trees, collecting for food banks, cleaning up beaches, and studying the size and activities of whales. All of these are activities that scouts might have done otherwise, but now their actions are informed with an understanding that what they do on the island can have a positive global impact. Environmental and community organizations like Beach Buddies have been recognized on the international stage for their impressive beach cleanup programs, which have engaged over 20 percent of the island's population in efforts to remove plastics from the beaches, as well as for outreach activities in the schools and community that are changing the mindset about litter and plastic and influencing government and community policy on plastic use. For example, MannVend has incorporated 100 percent compostable materials into all its vending and catering products, and financial-sector companies like KPMG have collectively committed to convert to 100 percent renewable electricity by 2030.

The Isle of Man government is continually assessing its policies with respect to the Biosphere philosophy. Even before its Biosphere designation, the Isle of Man had established extensive marine nature reserves by designating over 50 percent of the inshore waters and over 10 percent of the entire territorial sea as protected areas and committing to conserving species and habitats and enabling their recovery while also supporting sustainable fishing practices. And through the "Fishing for Litter" scheme, fishermen are now actively participating in the cleanup of marine litter. The government is committed to reaching net-zero carbon emissions by 2050, in line with the recommendations of the UN's Intergovernmental Panel on Climate Change, and because of the government's support, Rowan and her colleagues can move forward with their work to realize a plastic-free island and to develop a climate change response plan to achieve these goals. The island's Biosphere accreditation has aided this work and is providing a platform that showcases the results of these and many other innovative approaches.

The goal of all these initiatives is to create a sustainable future for the island. In addition, the culture and arts community, Culture Vannin (the Isle of Man's heritage foundation), has a paid Biosphere intern who is working on an exciting project called Mann's Green Footsteps, using innovative and impactful interviews, films, and other artistic projects to track the island's Biosphere journey. Since 2018, the Isle of Man Biosphere team has presented annual awards to organizations and individuals working to improve energy, economic, environmental, and educational sustainability. One of the 2018 winners was Zurich International, a global insurance company based on the island. They received the award for their commitment to integrating sustainable strategies across their entire operation. Within their core business, they have eliminated single-use plastics and disposable cups in the office, purchased office e-bikes for local transportation, and significantly decreased their print requirements. Zurich International is hosting employee webinars on climate change strategy and providing employees with carbon footprint calculators for their personal use. They also are involved in community initiatives, hosting company-led "Zurich Forest" tree plantings and beach cleanups and supporting local charities. I love this simple statement from a video the company made about its Biosphere accomplishments: "Our Island. Our Planet. Our Responsibility." Perfect.

For the people of the Isle of Man, the UNESCO Biosphere philosophy has been about how they manage everything about the island that they know makes it a special place. It's about doing not only what they can, but what's necessary to keep their landscape, nature, culture, heritage, and economy thriving. This is what all of us can be doing, wherever we live. We can make the necessary changes in our own lives and be part of a movement to encourage our crewmates to make necessary changes in their lives too, so that our actions spread across the entirety of Spaceship Earth.

My hope is that when our son Roman brings his children to the Isle of Man, and future generations bring their children too, they will respond to the place with the same awe and wonder I felt on my first visits. I'm hopeful that they will find the Isle of Man to be a green and sustainable place that's still thriving in harmony with its nature, and that the Biosphere philosophy that's been adapted so well to the island will have spread across our entire planet.

After speaking with Rowan, it became clear to me that the progress happening on the Isle of Man is a great example of how the success we've experienced on the ISS can be achieved by all right here on Earth. I saw similarities to how we live and work together on the ISS in the way residents of the Isle of Man approach life as an island community that's part of a bigger planet. It only makes sense that we can learn from their example and improve the way we live together on our planet as a whole.

While sailing to the Great Pacific Garbage Patch isn't an option for most of us, we each can take inspiration from Rowan's eXXpeditionary experience and her commitment to sharing it in a meaningful and actionable way. What she saw and learned led her to make changes in her own life and to become a change agent in her community. She began by behaving like a crew member of Spaceship Earth, and each one of us can make similar changes.

The ISS is the spaceship home of seven international crew members circling our planet every ninety minutes, and the Isle of Man is a small island home for roughly eighty-five thousand people in the middle of the Irish Sea, yet the work being done in both of these extraordinary places gives me hope that the same is possible for the more than seven billion of us who are traveling together under the blanket of the thin blue line on our planetary island in space.

On the ISS, it is absolutely necessary that each crew member consider how their actions will affect all the crew members and the spaceship's environment on which they all depend for survival. Rowan discovered that the same was true on her sailing voyage in the North Pacific. The people of the Isle of Man know the same is true for their survival in the middle of the Irish Sea. We must do the same on Earth. Everyone on Earth must come to this same realization. It will take each of us changing the way we behave, the way we live our lives. We all have to be willing to live differently, to do what's necessary in order to survive. We do it on a space station; we do it on the Isle of Man. We can do it on Spaceship Earth too.

Painting the first watercolor in space while on board the International Space Station during the Expedition 21 mission in 2009.

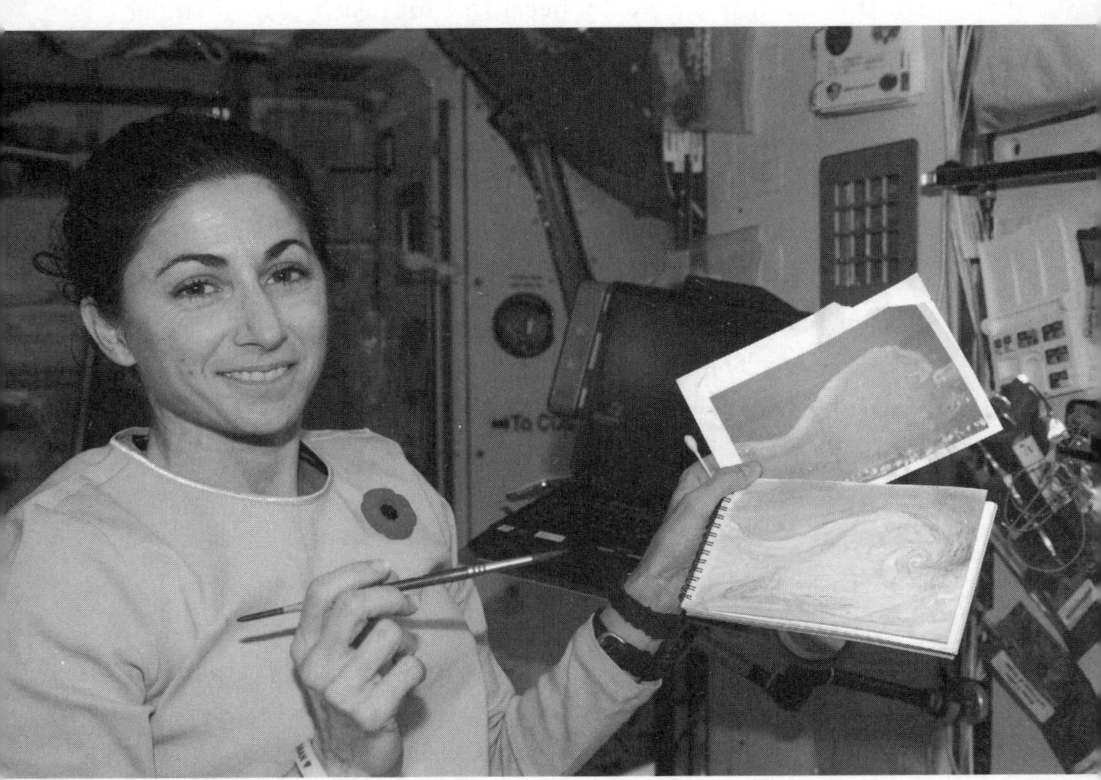

NASA

CHAPTER 3

LIVE LIKE CREW, NOT LIKE A PASSENGER

THE COOLEST THING about living in microgravity is the ability to float around, but perhaps the most entrancing aspect is to experience the magic of floating water. Astronauts have a tradition of taking what you might call a "water selfie"—a self-portrait of each astronaut's face, reflected in a floating ball of water. The face comes out upside-down in the picture, and it's a fun keepsake that enables us to share a little of what it's like to live in space.

Even bathing is fun. Because there's no running water on the space station, a "shower" consists of gently squeezing hot water out of the straw of a drink bag directly onto your body and watching it coat you like a second skin. Or you could squeeze out a softball-sized ball of water and stick an arm through it, creating a glove of water.

When I speak at schools, I joke with kids that if you squeezed out a big enough ball of water, you could float your whole body in and out of it. Though it's tempting to try that, we don't because it would be not only messy but dangerous with all the electronic equipment around. We do, however, do silly astronaut tricks, including placing several gummy bears in a floating ball of water, watching it spin around, and then slurping it all up in one giant gulp.

Just as when I'm here on Earth, the places I was most attracted to as I flew in space were the Earth's tropical regions. The stretch of coral reefs and shallow waters and little islands from the southern tip of Florida to the northern coast of Venezuela was my favorite. One night, as I was captivated again by the iridescent mix of blues and turquoises that swirled below, I lost track of time as I floated above the window with my camera and a large zoom lens aimed toward Earth. I snapped photo after photo in an effort to capture images of the ocean's beautiful, delicate patterns. Finally, I noticed it had gotten late, so I saved the images to my computer and went to bed.

The next day, curious about what I'd shot, I did some research and discovered that the spot I'd photographed is called Los Roques. I scrolled through my pictures and found one that captivated my imagination. In the photo, it appeared as though someone had reached down with a paintbrush and painted a wave on the surface of the ocean. I decided to make a painting of it. I printed out the picture of the wave on a scrap of paper and Velcro-ed it to the wall next to my sleep compartment. I gathered everything I would need to start my painting. Watercolor kit—check; paper—check; paintbrush—check; water bag—check.

I had to change my approach to painting, since traveling around the planet at five miles per second makes it impossible to paint what you're seeing out the window. But the experience of floating water changed my painting approach even more. The lack of gravity when you're floating in space adds a delightful layer of complexity and adventure to everything, including painting with watercolors. Since *everything* floats, I knew it would be important to have a plan for keeping it all organized. So I made sure each item had a little piece of Velcro on it, and I secured each piece to a wall or to the strips of Velcro on my pants, which we used to keep whatever we needed handy at all times.

I slid my feet under the handrails to secure myself. (I know that sounds funny, but we used "hand" rails both to pull ourselves along through the station and also to slide a foot under for added stability or to stay in one place.)

There's no dipping your brush into a cup of water in space, because there are no cups of water, so I slowly squirted a tiny ball of water out of the straw of my drink bag and then dipped the tip of my brush into it.

What happened next surprised me. I thought it was interesting enough to dip my brush into floating balls of water, but the way the water behaved with the tip of the brush and the paint and the paper was completely unexpected. Before I could bring my brush all the way to the ball of water, the water seemed to be attracted to the brush and moved toward it. The same thing happened when I directed the ball of water floating on the end of the brush over to the paint: it moved from the brush to the paint even before I touched it. The same thing happened with the colored water returning back to the brush. It was like some kind of mysterious magnetism was orchestrating the whole thing.

Beyond that novelty, I quickly discovered that if I were to have any success painting with a brush in space, I needed to change my technique because when the tip of the brush touched the paper, all of the colored water moved to the paper at once and became a big blotch of color.

"Dang it!" I had to start over. Several times.

After a few bouts of trial and error, I finally figured out that the trick was to use the brush to drag the colored water across the paper rather than touching the brush to the paper, as I was used to doing.

The image of Earth viewed from space as a glowing ball of blue floating against a backdrop of blackness reminded me how special it was to be living in such a unique environment, and the image became the inspiration for my painting. Like Earth-gazing through the window of the space station, painting in space was a transcendent experience. I experienced the physical and emotional sensation of floating while I painted, and I was able to immerse myself completely in the creation of my own interpretation of the view of our home planet from that very special vantage point. The painting itself is no masterpiece, but I'm happy that it's at least reminiscent of that awesome view.

Each time I painted, I had a chance to get absorbed in an activity that I love and at the same time to reflect on how thankful I was to even be in that place. I found myself thinking of what was in the plan ahead for the mission. On one such night, as a drop of water floated in front of me, I thought about our three new crewmates who would be joining us on the space station the next day.

On October 2, 2009, I watched the Russian Soyuz spaceship carrying our new crewmates, Max, Jeff, and Guy, hurtling through space toward us and then, in what seemed like slow motion, dock at one of our ports on the ISS. After a couple of hours of monitoring to ensure there were no leaks between the two spacecraft, we opened the hatches. Our friends floated into the space station, which probably seemed as wide and open as the Grand Canyon compared to the tiny spaceship they had just traveled in.

Our floating friends wore smiles of awe and wonder. (It doesn't matter how many times you've flown in space—the smile is the same.) Max Suraev, a Russian cosmonaut and a dear friend (he and his wife Anna are one of two sets of godparents to my son), was on his first flight to space and to the ISS; being on my first trip to space myself, I felt like I could totally relate to his smile and what he was feeling as he floated into the ISS for the first time. For my NASA astronaut colleague and friend Jeff Williams, this was his third flight to the ISS. I watched him smiling and reacquainting himself with the station, and I was looking forward to learning from his experience. Also sporting one of those wide grins, enhanced by a bright red clown nose, was Cirque du Soleil cofounder Guy Laliberté. He had paid the Russian Space Agency $35 million to train and fly as a "spaceflight participant" (aka space tourist). Guy was the eighth person to fly to space through this program.

While he enjoyed his time in space, Guy, like all astronauts, wasn't there just for the adventure. He had a greater mission.

"There was never any doubt that, for me, travel to space had to be meaningful and needed to serve one clear purpose. I wanted to raise awareness about water issues, and I needed to get the message out to everyone on the planet," Guy told me. "I can say that, no matter the project I take on, it's never a matter of doing something just for the sake of accomplishing it. I believe we have to put a little of ourselves into everything we do—a small bit of heart and purpose."

In 2007 Guy and his Cirque du Soleil team were starting to plan for their twenty-fifth anniversary, slated to occur in 2009 (the year Guy was also planning to take his spaceflight). They wanted to celebrate Cirque du Soleil in a way that wasn't just about touting their international success, but to use the event as a way to "give back" by highlighting their reason for forming the group in the first place: to "look forward and pursue our dream of a better world that has always inspired us." In researching ways to give back to make the world better, Guy quickly discovered that almost every single issue facing the world—poverty, ill health, poor education, social injustice, lack of economic development, environmental issues, gender issues, food security—had one feature in common: lack of clean water and proper sanitation. "Water touched me and inspired me," Laliberté said from space.

"Water is a source of life. When I learned a few years back that a child dies every eight seconds because of contaminated water I knew it was urgent to act."[1] He committed himself to doing something about it. In 2007, Guy formed the nonprofit One Drop Foundation to "ensure that clean water is accessible to all, today and forever." Along with the work of other key players in the sector, One Drop has had a role in the orders-of-magnitude improvement in that sobering statistic. The driving force behind Guy's spaceflight was his desire to further One Drop's mission.

Guy declared that his spaceflight was a "poetic social mission." In a preflight interview with *Wired* magazine, he said, "I have been described as many things throughout my 25 years with Cirque du Soleil. Fire-breather, entrepreneur, street smart, creative. I am honored and humbled today with my new job description: humanitarian space explorer."[2]

He went on to describe his artistic approach to his mission, "The purpose is clear: I decided to use this privilege [of spaceflight] to raise awareness of water issues to humankind on planet Earth. My mission is dedicated to making a difference on this vital resource by using what I know best: artistry. I am an artist, not a scientist, so it was my duty to contribute in my own way. I believe that with true art and emotion we can convey an important message. My wish is to touch people through an artistic approach and if we manage to do so, we will go beyond awareness."

His "poetic social mission" culminated in a two-hour live global "artistic happening" that took place on Earth and in space and that he orchestrated while floating 250 miles above the planet on the ISS. Called "Moving Stars and Earth for Water," the event featured musicians, including the rock band U2, former US vice president Al Gore, Cirque du Soleil performances, and other astronauts and artists in cities in fourteen nations on five continents, all coming together to highlight the role of water as "an inspiration and as a source of life" and to raise awareness of the need to protect and share this precious resource. My crewmates and I floated off-camera and watched as Guy joyously and effortlessly performed as emcee for the event. He shared videos of Earth that he had captured just days before through the space station windows, as well as stories of his experience in space, and he introduced the different participants from around the world. In each city at least one person would read from a poem by Man Booker Prize winner Yann Martel called "What the Drop of Water Had to Say," which is a poetic tale that tied together all the participants around the world. As a crew, we also had a role. Guy kindly included us in conversations with U2's Bono from Tampa, Florida, and with Julie Payette, our friend and a Canadian astronaut participating from Canada. Everyone involved focused on our very special vantage point for appreciating the beauty and fragility of our water planet.

As a result of creative, large-scale events like this one and like the many other "novel" fundraising efforts by the organization ever since its inception, One Drop has not only raised global awareness of water issues but implemented sustainable international projects that support the UN-developed WASH (Water, Sanitation, and Hygiene) programs, including its own unique Social Art of Behavior Change (SABC) programs. As of December 2018, One Drop projects had provided access to clean water and safe sanitation, as well as education on healthy behavior, to more than two million people in thirteen countries. All aspects of the One Drop programs utilize what they describe as "social art"; the "brainchild of Guy Laliberté," social art is based on creating a sense of involvement, ownership, and empowerment in the communities One Drop works with while at the same time approaching the water challenge through the artistry that has defined Laliberté's career.[3]

"The approach that has been developed at One Drop is centered on the arts," Guy also told me, "and is profoundly linked to people's emotional response, but at the same time, it is completely evidence-based and results-oriented. And it works. When you reach someone's emotions, they (and we) can accomplish the seemingly impossible."

I love the way Guy speaks of his space mission because he clearly understands and conveys that all the work we're doing in space is about improving life on Earth. From day one, he saw his mission as another way to demonstrate that we all can work together in support of a common and challenging goal, and that together we can "make a difference, one drop at a time."

On the space station, we have a saying that "yesterday's coffee becomes this morning's coffee." As on Earth, water is a precious and limited resource in space, and it's heavy, so it's very expensive to resupply from Earth to space. This is one reason why we do things like send our food to space dehydrated, so that it weighs less; that enables us to get more at one time on the resupply missions. To the shock of many, though, one of the primary ways we manage this precious resource in space is to recycle water from every available source, including crew members' own urine and sweat. The ISS system, known as the Water Reclamation System (WRS), can recycle about 90 percent of the mix of wastewater liquids it receives into clean drinking water through a three-step process: (1) particles and debris are first filtered out; (2) distillation and more filtration removes organic and inorganic impurities; and finally, (3) a "catalytic oxidation reactor" kills bacteria and viruses and removes volatile organic compounds (VOCs—yep, the same kind that are removed from the air). This conversion of urine to clean drinking water might sound disgusting, but the water we drink on the ISS is purer than the water most people drink at home on Earth.

The entire crew of the ISS shares the responsibility for sustaining, managing, and restoring vital resources like water and preserving the integrity of our life support systems. Likewise, even though we might not acknowledge it, each of us on Spaceship Earth has the responsibility to sustain, manage, and restore Earth's vital resources and preserve the integrity of our life support systems here too.

One way we can do this is to increase our awareness of where the clean drinking water we take for granted comes from and how we can consume it responsibly. I was amazed at what I could learn by just searching my local state, city, and county websites. This awareness has led me and my family to change the ways we consume water and to be more active in our community and involved in the decisions being made on our behalf for the management and protection of our local water supply.

For example, in 2017 I moved to a city that offers the public reclaimed water to hydrate our lawns. We not only avoid the cost and waste of using drinking water to sprinkle our grass but also get the bonus of adding a natural fertilizer to the grass. This eliminates the need to use artificial fertilizers, thereby reducing harmful runoff flowing into the community's water system and the surrounding waterways. It also improves the quality of the water that's cycled through the Earth's natural process. These are significant benefits from one small change.

One of the reasons I was willing to strap myself to a rocket to space was that I understood that all the work and science performed on the ISS is ultimately about improving life on Earth. The same processes developed and used to provide clean drinking water on a space station are helping to improve access to and provision of clean drinking water here on Earth.

Space-based water technology has been developed commercially for water treatment in residential, business, and industrial applications, and it also has been deployed to provide humanitarian aid and disaster relief for communities in need worldwide. NASA and our international partner space agencies have continued to facilitate this type of development through their ongoing technology transfer and spin-off programs. These agencies work in partnership with companies and aid organizations to deploy water filtration and purification systems to populations in some of the most remote and water-challenged areas of the planet, as well as specially designed water bags that incorporate a compact form of this technology as first-response equipment for natural disasters, refugee camps, civil emergencies, and other remote locations.[4]

Aquaporin, mWater, and Bioclear are three companies that employ water technology developed for space to serve their customers.

Aquaporin is using advanced water filtration for freshwater treatment and desalination; mWater has developed a mobile phone application, based on the water quality monitoring and communication system developed for the space station, to manage water cleanliness in over 150 countries worldwide; and Bioclear has found uses for space technology in a variety of commercial applications, from tracking hygiene issues in drinking water to monitoring bacteria in contaminated soil.

In addition to the spin-offs to commercial and humanitarian uses of space technology, NASA itself brings this same technology back to Earth for its own use. At the NASA Ames Research Center in California, they have designed and built Sustainability Base, which is described as the "first sustainable space settlement" here on Earth.[5] The facility utilizes NASA innovations originally engineered for space travel and exploration within the fifty-thousand-square-foot, crescent-shaped Sustainability Base. The building is a working office space, a showcase for NASA technology, and an evolving example of sustainable architecture for the future of building design. Through a combination of NASA innovations and commercial technologies, Sustainability Base leaves virtually no footprint. NASA has made every attempt to apply the closed-loop thinking used for space exploration to erecting a green building on Earth.

Sustainability Base uses integrated data, information, and smart systems to "anticipate and react" to changes in sunlight, temperature, and usage, and the people who work there are able to self-monitor their energy usage as the building suggests ways to conserve responsively. Power is generated through a combination of solar and wind power and high-efficiency fuel cells (the latter originally designed for future use on Mars). The building materials and furnishings were locally procured and used many reclaimed or recycled materials, and the site itself is complemented by all-native landscaping. Even the shape and placement of the building and its floor-to-ceiling windows were designed to maximize daylight exposure, which provides occupants with power benefits as well as a visual and emotional connection to the surrounding landscape and a constant flow of fresh air. The occupants have described it as a workplace with "open and abundant connections to nature." By using the same water recycling systems that were designed for use on the International Space Station, Sustainability Base uses 90 percent less potable (drinking) water than a traditional building of comparable size.

When I reflect on the "through a spaceship window" view I had of the floating ball of blue that is our planet, I often find myself thinking about the popular quote "water water everywhere, nor any drop to drink," from Samuel Taylor Coleridge's *The Rime of the Ancient Mariner*, the poetic tale of a sailor stranded on a ship surrounded by all the water of the ocean, but without a drop to drink. In some ways we're a little like that sailor. Our planet is blue because about 75 percent of its surface is covered by water, but most of it is saltwater. Only 2.5 percent of water on Earth is freshwater, and nearly all of that water is frozen—locked up in polar ice caps, glaciers, and other ice. The small amount of freshwater that remains is all that's available to everyone on Earth and for all the ways we use it. NASA monitors Earth's water from space, the skies, ground stations on land, and ships sailing the seas, and even with mobile phone apps. "'All the water on Earth already exists. We can't make more,' said Bradley Doorn, program manager for NASA Earth Applied Sciences' Water Resources Program. 'We can only track it, predict it and protect it as it cycles around our world.'"[6]

Because liquid water, especially the scarce fresh kind, is an essential requirement for all life on Earth (right down to the cellular level), NASA tracks nearly every aspect of the water cycle—as it falls from clouds as precipitation, as it lies underground as groundwater, as it soaks into the soil, as it moves into rivers and lakes, as it's taken up by plants and used by animals and humans, and as it evaporates back into the atmosphere.[7] I was surprised to discover that about 70 percent of the limited resource of freshwater on our planet is used for agricultural irrigation (wow!). In addition to monitoring how water is used and its quality for drinking, the significant quantity of water in various locations (sometimes too much due to flooding or too little due to drought or overuse) is also monitored by NASA, in partnership with other international space agencies, such as JAXA (the Japan Aerospace Exploration Agency). Together they carry out important missions like the Global Precipitation Measurement (GPM) and the Tropical Rainfall Measurement (TRM), which use satellites to provide high-quality estimates of Earth's rain and snowfall every thirty minutes. The imagery and data from satellites like Landsat and the ISS-based instrumentation ECOsystem Spaceborne Thermal Radiometer on Space Station (ECOSTRESS) are vitally important for helping us track, understand, and better manage and protect water for the benefit of all life on the planet.[8]

All these examples of how we're utilizing space-based technology to bring what we do in space back to Earth show us just how effectively we can live like crew and adapt this technology for solutions to our planetary problems. This is what the ISS motto, "Off the Earth, For the Earth," is all about. Further, should we choose to model life on Earth after an ISS mission and the way the crew and ground team foster international cooperation, we could also be living "*On* the Earth, For the Earth," and for all the life it supports.

Sounds like an audacious vision, and it is. Oversimplified? Maybe. But each of us, you and me included, needs to make a personal decision about the role we will play on Spaceship Earth. Are you going to live like a passenger, or are you going to suit up and join the crew?

I have been so impressed by the positive impact that one person can make in the world when they experience the reality of a new perspective and are inspired to take action in response. When Guy Laliberté discovered how many people on the planet are suffering because of the lack of clean water—which up until that point had been invisible to him—he chose to do what he could to make things better.

"We must believe in our ability to influence the people around us through our behaviors, ideas, and projects.... It's a domino effect. Individuals who have single-handedly brought about change this way are countless," Guy said. "The immensity of space and the majesty of our planet cannot help but highlight how small each of us is, individually. And yet they inspire hope and a certainty that everything has significance, even beyond what we usually see. I believe that every action makes a difference, no matter how big or small or humble or bold it might be. We can all add our drop to the bucket."

Of course, you don't have to go (I'd even argue you shouldn't have to go) to outer space to live like a crew member on Earth and make a positive impact. Each of us can be one of those people adding a drop to the bucket.

In April 2020, I was fortunate to interview Scott Harrison, founder of "charity: water" and author of *Thirst: A Story of Redemption, Compassion, and a Mission to Bring Clean Water to the World.* At age twenty-eight, Scott had been a nightclub promoter in New York City for a decade, living as a self-described "selfish hedonist like I was the only one on the planet." In stark contrast, Scott's parents had raised him within a deep Christian faith. During those party years, when people would ask Scott's father how he was doing, his father would reply, "Please pray for my son, he's gone prodigal."

On what would be his last international party trip, instead of participating, Scott became an observer. And as he watched the debauchery, he had an epiphany of sorts. A thought began to dawn on him. "I had reached a really dark and depraved and emotionally and spiritually bankrupt point in my life," Scott said. "I asked myself, *What would the exact opposite of my life look like?*" To answer that question, he decided to commit one year of his life to service to others.

Scott's transition from a life fueled by selfish motivation to one based on empathy and compassion had begun.

The next question was: Who could he help, and how? Scott's reputation preceded him as his applications to serve were rejected by several humanitarian organizations. Finally, Mercy Ship, a floating hospital that brings medical help to those in need, allowed him to join one of its missions, but he had to pay his own way to be on the ship. Scott sold all his earthly possessions, and what started out as a "trade"—one year of service to make up for one of the years he'd selfishly wasted—turned into a two-year supporting role as a photojournalist for the organization.

Scott's first mission with the organization was to Liberia, one of the poorest countries in the world. During that mission, he saw for the first time what extreme poverty looked like. It was "the kind of poverty that has human beings drinking dirty water from brown, viscous swamps or ponds or rivers, and children regularly dying from water-borne diseases. It struck me hard that this wasn't okay," Scott said.

Profoundly affected by the pain and suffering he witnessed, Scott did the best he could at the job he was given. He photo-documented the story of every person who lined up for their chance at surgery on the hospital ship. He took the before-and-after pictures of patients with extreme conditions like leprosy and tumors and cleft palates. He began to promote these people's stories instead of promoting nightclubs. He kept this photo record for the Mercy Ship's own files, but he also shared the images with the network of people he had established back in New York City. People started to donate in support of the Mercy Ship's work. Scott realized that he could inspire positive action even from people he hadn't known for their compassionate or empathetic side. Between his tours with Mercy Ship, he went back to New York City, continued to share the story through his compelling photography, and raised even more money for the organization's work.

On his second trip to Africa, Scott was faced again with the shocking truth about the lack of clean water and its tragic consequences for so many people on the planet. Globally, about one in ten people do not have access to clean water. Everywhere he looked while in Liberia, he could see that almost every aspect of people's lives was affected by the lack of clean water.

Reflecting on the entire experience, Scott said: "My Earthrise moment was day three on that first trip to Africa. I was at a patient screening, and there were five thousand sick people standing in a parking lot waiting for us to open the doors, and all hoping to get a chance at fifteen hundred surgery slots. I'd never been confronted with any kind of mass or swarm of human suffering in aggregate like that before. Then realizing many of those people had walked for more than a month just with a hope of seeing a doctor. Then realizing the conditions that so many of them were living in. Realizing we had the only CT scan machine in four neighboring countries. I think it was just how far away the reality facing these people was from my own life's experience.... It was in such stark contrast to my former life, where I sold bottles of water at nightclubs for $10, that it led me to really want to work on water and want to bring clean drinking water to everyone alive. That led me to charity: water."

Scott formed charity: water in 2006, when he realized that he could help change the world for the better. Charity: water focuses on providing rural communities with their first access to clean water. As of 2019, the organization had raised over $400 million from donors, which, through its "100 percent goes to the field" model, has directly funded over fifty thousand water projects to bring clean water to more than eleven million people in twenty-nine developing countries around the world. These impressive numbers tell the story of how charity: water is working toward its goal of bringing clean water to the world, but I'm particularly impressed with the effective team the group has built to make this happen.

What started with Scott's personal mission to create change, with the help of a couple of passionate supporters working with him in a friend's borrowed New York City apartment, grew into a movement of positive change that has brought together a core team in the United States supported by an international team of partners in the field who have the expertise to implement the water projects. The effort is also assisted by over one million donors providing financial support and by local community representatives, who are integral to defining the best sustainable solution for the project located in their community. Other organizations, like Laliberté's One Drop, are also working with charity: water toward a shared goal of clean water for all. All these individuals are working together as a crew and cooperating through their different strengths to "provide clean water to the world."

I mentioned to Guy Laliberté the similarity I saw between our space station crew and a Cirque du Soleil performance. "I see in the well-choreographed and executed performances how the cast needs to bring their individual best and work together as a crew," I said. "I am in awe of how everyone manages to be in the right place at just the right time. Every aspect of success depends on this. If one person isn't where they need to be at the right time, then the whole performance will fail and people can get seriously hurt. Sounds like spaceflight. Sounds like how we should be behaving as Earthlings on our Spaceship Earth."

Guy agreed and added, "Just like a show from the Cirque du Soleil, we are all part of a troupe that must deliver the performance of a lifetime. We must work together, learn to trust each other, go beyond our limits. If one falls, we all fall. In order to ensure our success, the concentration of each person is vital, and no individual must be left behind. The survival of our world will depend on the choices we make today and will be decisive for the future of the next generations. Inevitably, the journey will be strewn with pitfalls that will sometimes make us doubt our abilities. But the show must go on."

As astronauts, we train for years to participate in the ISS mission of science and international partnership, where we bring together a team of people representing fifteen different countries, with diverse talents and expertise, all working together to contribute to one mission—"Off the Earth, For the Earth." The most important part of the training is learning how to work as part of a crew with a shared mission, and especially how to work together when things don't go as planned.

So much of astronauts' training involves what they discover about themselves outside their comfort zone. We trained in a simulator, in a big swimming pool, while flying in a T38 jet, and as we struggled through a Russian language class. Sometimes we were learning sea survival in the Black Sea, or cold weather survival in the frozen wilderness of Canada or Russia. (My crew trained during the coldest Moscow winter in over a hundred years.) We trained for eighteen days while living and working on an "inner space" mission in an undersea laboratory sixty feet beneath the surface off the coast of Key Largo, Florida. These places and experiences took us out of our comfort zones and gave us the opportunity to discover a lot about our own strengths and weaknesses, as well as those of our crewmates. Just as important, we learned how we could best utilize our mix of strengths to overcome the challenges presented to us.

On a space station, we accept the reality that we're living 250 miles above the Earth, with only a thin metal hull between us and the deadly vacuum of space. This situation would naturally be at least a little out of anyone's comfort zone. So all the training we do beforehand, together as a crew and individually to become better crew members, is what allows us to be there and not live in fear of what lies beyond that thin metal hull, as well as to be as prepared as humanly possible to deal with challenges as they arise. We know we have to rely on one another, that every crew member must be doing their job, not only for the mission to be a success, but also, more importantly, for the crew to survive. The best crews—both in space and on Earth—are the ones you know you will enjoy sharing the experience with; after all, time in space is limited and should be enjoyed. The best crews are also the ones you know will have your back, and will trust that you'll have theirs when things don't go as planned.

One of the best representations of a crew appreciating the unique experience of spaceflight, but also understanding the implications of just being in that environment at all, is a crew picture we took during Guy's time with us on the space station. I show this picture to every group of people I speak to, especially kids. For me it is such a beautiful and fun picture of our crew. All nine of us are in the picture wearing our blue "dress" flight suits, each with our country's flag attached to the left shoulder and, more importantly, our crew mission patch worn on the chest. We are floating in all different directions (there's no better way to show that you're in space) and wearing big smiles and red clown noses (a gift to each of us from Guy), and we are all flashing a peace sign. Though some might think it looks unprofessional, what I see is the personality and the professionalism of each member of our crew shining through. Every crew in space takes many crew pictures together, but I still think this is one of the best; for me it epitomizes one of my favorite phrases: "putting the human in human spaceflight."

As Earthlings, Earthrise is *our* crew picture. It represents the who and where we all are in space together. It represents the beauty and fragility of our planet. It represents a paradise and a place in peril. It represents a place where we can experience the greatest joys and the greatest suffering. It represents our shared life support system; for all our exploration, we have found no other like it in all that blackness that surrounds us. It represents a sense of love and connection and the need for us to respect and care for our planet and one another. It represents the necessity for all of us to rise together and work as crew on Spaceship Earth.

Here on Earth, we are being presented with some daunting challenges, and I'm pretty sure that most people around the world, as individuals, are feeling at least a little out of our comfort zones. Guy and Scott undoubtedly both found themselves out of their comfort zones as they were forming their organizations, and they probably continue to find new challenges as they forge ahead.

Guy was profoundly impacted by the truth he discovered in the human casualty rates associated with dirty water, and Scott found himself physically immersed in the uncomfortable reality of that statistic when he was standing in front of five thousand people hoping for one of fifteen hundred surgery slots, and when he witnessed children who were sick and dying from drinking dirty water. They were both confronted, in a very uncomfortable way, with one of our most challenging planetary problems: how do we provide all of humanity with a resource, clean water, that is so basic to survival?

The action they felt compelled to take in response to this new awareness is a powerful reminder for us all of the difference that one person can make. While it's natural to feel like we couldn't possibly have a positive impact on the planetary challenges we face, we can take inspiration from their stories. I don't know if I'll ever have the opportunity to form an organization like One Drop or charity: water, but I can choose to support them. I can make a choice to discover the reality that exists around me, to dig deep and learn more about it, and to find a way to take my own personal action. I can choose to apply my own time and talents. I can choose to be an active part of the solution.

Harkening back to Scott Harrison's father's request that family and friends pray for Scott because he'd "gone prodigal," I recall being struck by these words when I first read them. I knew these were the words of a father who was worried for his son and seeking the support of others to lift him up in prayer, to help bring him "home." I remember smiling.

I think the reason these words struck me, and why I smiled, is that in one way or another, it occurred to me, we all have gone prodigal somewhere along the way. We all have at one time or another been in need of redemption. Perhaps we've wandered from home by the way we've separated ourselves from understanding our interconnectivity with Mother Nature. We may have become apathetic to our planet's life-giving nature, and as a result we need to find our own prodigal shift, from indifference to empathy and action. Scott's father's prayers were answered. Perhaps our own redemption—our prodigal journey back home—can come about from committing ourselves to setting aside living life as a passenger and beginning to take personal action to live like a crewmate on Spaceship Earth.

My NASA astronaut class was nicknamed "the Bugs." On the International Space Station, my STS133 Bug crewmates Eric Boe, Mike Barratt, Al Drew, Steve Bowen, and I are holding a quilt made by our fellow Bug Karen Nyberg.

NASA

CHAPTER 4

NEVER UNDERESTIMATE THE IMPORTANCE OF BUGS

ON THE EVENING OF JULY 20, 1969, I sat in front of a black-and-white TV in the living room of my childhood home, eating a grilled cheese sandwich as I watched NASA astronauts Neil Armstrong and Buzz Aldrin land the lunar module *Eagle* and take their first steps on the Moon. Even at age six, I knew that this was an extraordinary event.

Quite a few astronauts credit this Moon landing moment for their lifelong goal to become an astronaut. I did not have the same reaction. Although I had flown in airplanes since before I could remember, I'd always thought that being an astronaut was something only "special people" got to do. Even so, I enjoyed the perspective of being way up high in an airplane and looking down on all the cars and houses that seemed so large and far away on the ground, yet looked like a miniaturized toy land from an airplane.

My first memories of flying were as a passenger in my dad's home-built biplanes, which had names like *Starduster2* and *Skybolt*. My dad was a businessman, and he loved to build and fly small airplanes. As a family, we spent a lot of time at the local airport. Because my dad shared what he loved with me, I discovered my own love of flying—of separating from Earth and experiencing it from above—and my desire to know how things fly.

I received a similar benefit from my mom, who nurtured all kinds of interests, including my artistic side. My mom is a nurse and a very creative person. When we were young, most of the clothes my sisters and I wore were sewn by her. It was the '60s and '70s, so she did macramé and pottery and made hooked rugs, and she invited us to join in. Mom also felt it was important for me to be well rounded physically, so she made sure I got enrolled in ballet classes and joined the community softball team.

My mom was the one who made sure that my sisters and I spent time at the airport—the Clearwater Airpark in my hometown of Clearwater, Florida. The airport had an apt name because to us kids, it felt a lot more like a park than an airfield, and we found all kinds of adventures there and things to explore. It had one grass-strip runway, a couple of rows of metal hangars, some covered parking spots for the airplanes, and a small trailer for the office, all surrounded by wide-open fields and small patches of trees with picnic benches.

My favorite part was flying in airplanes, but most of our time there was spent on the ground. Besides my parents, other adults brought their children to the airpark. We kids would play while our parents worked on airplane projects with their friends or socialized, which mostly meant talking about airplanes over coffee and cigarettes (very much in fashion during the 1970s) and some flying.

During the "not flying" time, we kids looked for other ways to have fun. When I was around the age of seven, my main adventure, along with my sister Shelly, was watching and chasing damselflies. Slender and delicate, with little segmented and sticklike bodies and round, bulbous eyes, most of them were dark with clear translucent wings, but every now and then we'd see some turquoise, shiny emerald, or purple and blue damselflies with dark or multicolored wings. They all looked ethereal, like little Tinkerbell fairies dancing around the airfield. We didn't know what they were, so we called them dragonflies. The adults would point out that "those aren't dragonflies," but they didn't know their proper name either, so we kept calling them dragonflies.

One big difference we discovered between damselflies and dragonflies was that damselflies didn't bite. Every time we went, it seemed like there were thousands of damselflies swarming around us, and I wanted to be right in the middle of them. I figured out that if I stood really still with my hand out, sometimes one would land on the end of one of my fingers. I remember once staring into a damselfly's cute little bug eyes and thinking that they looked like a tiny pair of flight goggles. The damselfly appeared to be staring back, as if it were trying to get a closer look at me too. Some days I looked forward to seeing the bugs even more than the airplanes.

As I reflect on my memories of the damselflies, I realize I had not considered them in decades. They faded from my awareness as I grew up, and as my interests shifted away from chasing bugs that fly to learning how to fly myself. I spent less time at the airpark after I went to college, and the grassy runways were eventually paved over. The joy I once experienced with the damselflies faded without my even noticing.

How could something that held such a vivid and special place in my childhood have disappeared from my consciousness like that? Maybe, as I got older, when I was at the airpark I became more involved and interested in how the airplanes flew. Maybe I took it for granted that the damselflies were there doing their damselfly jobs. To me, the damselflies had always been there, so I'd assumed that they always would be.

It wasn't until I was writing this chapter and remembering the fun I'd had with the damselflies that I realized I couldn't recall the last time I'd seen one anywhere—not even at the Clearwater Airpark. I'm happy to report that, as of this writing, it appears that they are still alive on this Earth. Still, this realization felt like a wake-up call for me. Even though I'm told that I still share the planet with damselflies, the fact that now, when I look for them in places where I know they thrived in the past but discover they are no longer there, it is shocking and sad to me. I intend to use this case of the "missing" damselflies to remind myself to look more closely, to pay attention to the environment around me, to slow down enough to become aware not only of what I see but of what I'm not seeing anymore, and to begin to ask why.

One week before my sixteenth birthday, my dad was killed when his small airplane crashed. Although losing him was devastating, I had grown up with his love for flying. When he went out that day, it was something he had looked forward to. We were building a house on a lake so that my dad could land the new amphibious airplane he was building on the water in our own backyard. My mom finished construction of this house, and I lived there until I went to college. My mom lived there until she sold the house in 1997.

My dad's accident did not dampen my interest in flying. If anything, it deepened my desire to learn about airplanes—to know more about how they fly—and to understand what might have caused the crash.

In this pursuit of flying and wanting to know more about how airplanes fly, I am fortunate to have had a lot of female role models. One of the other people besides my dad that I would fly with at the local airport was a young woman named Val Spies, who appeared to be in her midtwenties. She and her husband ran the glider operation at the airport, and she would take me flying in her Cub. (She also taught me how to drive a car with a manual transmission on a hill.) I had lost touch with Val over the years, but we reconnected when I was getting ready for my first spaceflight. These days she runs a beautiful yoga studio in Tampa, Florida.

In high school I had several female teachers, including Mrs. (Helen) Wilcox and Mrs. (Marion) Steele, who encouraged me to study more challenging subjects than I would have chosen on my own. Although I never focused in my career on biology, their favorite subject, the work I did in their classrooms (among students I'd always thought of as way smarter than me) was a great lesson in the value of challenging myself to learn new things.

My favorite class in high school was an elective called Intro to Aviation, taught by Mrs. (Elizabeth "Dusty") Ransom. I was so excited to discover that my school offered it. Who knew that you could go to high school and take a class about something you already knew you loved? And on top of that, it was taught by a woman who had done things I'd only read about! She never boasted about any of the amazing flying experiences she'd had, but all of us in the class wanted to know more about her background. She'd earned her pilot's license at age nineteen at the University of Toledo. At age twenty-two, she had helped the World War II effort by joining the Women's Airforce Service Pilots (WASP), which ferried bomber aircraft around the country. She later earned her bachelor's and master's degrees in education. A woman of many talents, Mrs. Ransom also was a swim instructor, a master gardener, and a champion quilter. I was fortunate that she'd landed in 1968 at Clearwater High School, where she taught aviation science, anthropology, and history; she remained there until her retirement in 1986, six years after I graduated.

Mrs. Ransom passed away in 1997. While I'm sorry I never had the opportunity to meet up with her again after high school, I'm thankful that shortly after I'd been selected as an astronaut, I got to meet with her daughter at the same Clearwater airport where I grew up and tell her how impactful her mom's aviation class had been for me. At the time, I never considered that there was any significance to having female aviators and scientists as role models to encourage me—they were just people in my life who shared with me what they loved. But now I realize that the presence of these women in my life helped me discover my natural interests and aptitudes, which led to my career as an astronaut.

The greatest female role model of all is, and always has been, my mom. She was thirty-nine with three daughters when my dad died. (I was fifteen, Shelly was fourteen, and Noelle was eight.) I am in awe to this day of her strength and commitment in encouraging me to fulfill my dreams. She never once asked me not to fly, or told me that I couldn't fly, or even suggested that I shouldn't. She supported me every step of the way as I decided I would earn my pilot's license right out of high school and then study aeronautical engineering at university, and she stood strong, watching with my sisters, as I climbed aboard rocket ships and launched twice into space.

I knew all along that it is a lot harder to watch someone you love strap into a rocket than it is to be the one strapping in, and now that I'm a mother too, I know that particular angst of watching your child do something potentially dangerous. But I didn't fully appreciate what it might feel like for a child to watch a parent take a huge risk until, strapped in and ready to take off for outer space, I thought of my son, who was just seven years old and standing on the roof of the launch control center. What courage it took for him to watch the Space Shuttle take off with his mom inside.

From the time I watched the first Moon landing, I thought being an astronaut must be an extraordinary thing, but it was a long time before the idea of going to outer space seemed like something that I could, or even should, consider. But eventually, in the year 2000, I was selected as a member of the eighteenth class of NASA astronauts. It was a great honor (and honestly a huge surprise), as only seventeen of us were chosen from over five thousand applicants. What surprised me most in being chosen was not the odds against it, but the fact that I'd applied at all.

I'd studied aeronautical engineering at Embry Riddle Aeronautical University (ERAU) in Daytona Beach, Florida, which is only fifty miles north of the Kennedy Space Center. The campus made for a pretty awesome place to watch Space Shuttle launches. While studying to learn how airplanes fly, I discovered I also wanted to know how rocket ships fly. This led me to not just watching the launches at the KSC but also pursuing a job there with NASA. In 1988, when the

Space Shuttle program was returning to flights after the *Challenger* accident, I started work there as a NASA engineer. I worked at KSC for almost ten years before it even occurred to me to apply for astronaut selection. As I was helping prepare Space Shuttles for launch, I had gotten to know what astronauts do during the 99.9 percent of the time when they're not flying in space, and I realized that at least 80 percent of their job was a lot like what I was doing as a NASA engineer. That's when I spoke to some of my mentors and they encouraged me to pick up a pen and fill out the application.

"Pick up the pen and fill out the application?!" How could something so simple seem so daunting? For weeks I asked myself the same "why would they ever pick me?" question. I almost self-doubted my way out of an amazing opportunity, had I not finally taken the one action in the whole process that I had total control of—filling out the application. I am so thankful that no one I spoke to discouraged me. The simple encouragement from everyone I spoke with to "pick up the pen" gave me the confidence to do so, and that application would open up a whole new flying adventure and a whole new perspective on life.

Per tradition, the class before us was tasked with giving our class a nickname, and they christened us "the Bugs." That name was inspired by the "millennium bug"—the widespread concern in 1999 that computers would crash and end life as we knew it when calendars rolled over from 1999 to 2000. While intended to be playfully disparaging, the name could have been worse. Our predecessors had received names such as the Maggots, the Hairballs, the Penguins, the Sardines, and the Flying Escargot. We embraced ours; after all, most bugs have wings and can fly. The one thing all astronauts want to do is fly.

Since I've become an astronaut, and even before I flew in space, the question I've been asked most often is, "How do you become an astronaut?" The truth is, not one of the astronauts in my class got there the same way. The one thing we all had in common was that we loved what we were doing before becoming an astronaut. We were passionate about the work we were doing on Earth and curious to learn more, we all were looking for ways to use our innate curiosity and commitment to work to help improve life on Earth, and we all discovered at some point that becoming an astronaut might be another way for us to do that. To winnow down the applicant pool (there were over eighteen thousand applicants for the 2017 class of eleven), NASA has basic requirements for education, work experience, and general health, but otherwise the qualifications are pretty wide open. While restrictive in some ways, these criteria still make it possible for all kinds of people from many different backgrounds and experiences—even someone like me—to become an astronaut.

Though many companies still don't avail themselves of the advantages of a diverse workforce, the merits of both social diversity (race, gender, ethnicity) and diversity of background and expertise in the workplace have been demonstrated across all industries, and it has certainly been demonstrated for the crews of space program missions. One of my favorite examples is the ISS program.

The fact that five international space agencies and astronauts representing fifteen countries work together on the ISS program has been an exemplary model of cooperation across borders and cultures for all of us on Earth. Not only has the ISS program demonstrated that leveraging multinational ingenuity and diverse expertise is more effective technically, but it has also strengthened relations among the partnering nations and made research in space more affordable. Even when there's tension between participant nations on Earth, the cooperation on the ISS continues.

While our Bug class of 2000 was not particularly remarkable with regard to diversity of gender and race, the diversity of expertise among us certainly was. We were made up of engineers of all kinds, military test and fighter pilots, medical doctors, an oceanographer, a geophysicist, and even a submariner. Each member had their own outside interests that complemented our combined experiences, such as race car driving, sewing, painting, semiprofessional water skiing, woodworking, flying, gardening, ballroom dancing, cooking, rock climbing, music, house building, and scuba diving, to name a few of the multitude of activities the Bugs enjoy. One thing we discovered—and that NASA knows to be true—was that our varied individual backgrounds and experiences made us a stronger team.

Throughout its history, NASA as an agency has recognized the value of both social diversity and diversity of expertise and experience within its human spaceflight programs. Although there are way fewer women and people of color among university graduates in the STEM fields and industries that are feeders for positions at NASA (including astronauts), NASA itself has been proactive and progressive in recruiting.

Considering all that had to happen to land humans on the Moon in 1969, it's clear that one of the most visible events that launched it all was President John Kennedy's historic speech at Rice University Stadium in Houston, Texas, on September 12, 1962, when he said: "We choose to go to the Moon in this decade and do the other things, not because they are easy, but because they are hard." This presidential directive to go to the Moon (and safely return) was delivered two months before I was born—and less than seven years later, I watched it happen live on the television in my family's living room.

What President Kennedy said leading up to this statement is also inspiring:

> The exploration of space will go ahead, whether we join in it or not, and it is one of the great adventures of all time, and no nation which expects to be the leader of other nations can expect to stay behind in the race for space.... We mean to be a part of it—we mean to lead it.... In short, our leadership in science and in industry, our hopes for peace and security, our obligations to ourselves as well as others, all require us to make this effort, to solve these mysteries, to solve them for the good of all men, and to become the world's leading space-faring nation.... We set sail on this new sea because there is new knowledge to be gained, and new rights to be won, and they must be won and used for the progress of all people.... There is no strife, no prejudice, no national conflict in outer space as yet.... Its hazards are hostile to us all. Its conquest deserves the best of all mankind, and its opportunity for peaceful cooperation may never come again.... We choose to go to the Moon. We choose to go to the Moon in this decade and do the other things, not because they are easy, but because they are hard.

Why am I so inspired every time I listen to or read these words from President Kennedy's speech? Because he was speaking not just about going to the Moon, but about the need for us to continue to challenge ourselves to be the best, to be leaders, to make what seems impossible possible. His words are about hope and the benefit of space exploration for the progress of all people.

Less than a year later, in 1963, President Kennedy made another speech, this time calling for Congress to enact legislation "giving all Americans the right to be served in facilities which are open to the public," as well as "greater protection for the right to vote."

This call was answered at NASA, interestingly, by Dr. Wernher von Braun, a German scientist who developed the V-2 rocket weapon used by the Nazis and who had immigrated to the United States after the end of World War II. He later applied that same technology in designing the Apollo Saturn V launch vehicle. In 1964, Dr. von Braun, who was then the director of NASA's Marshall Space Flight Center in Huntsville, Alabama, addressed the Association of Huntsville Area Companies about the need to recognize the civil rights of all Americans and to work against discrimination. "According to Fred Schultz of General Electric's Space Division [one of the companies in attendance and a contractor for NASA], von Braun's remarks gave association members 'the backing they needed to launch further successful drives for equal employment opportunities.' "[1]

President Lyndon Johnson, after whom NASA's Johnson Space Center in Houston is named, also pressed for landmark legislation that outlawed discrimination. He signed the Civil Rights Act on July 2, 1964, the same day it was passed by Congress.

The Apollo program was, in part, a response to competition with the Soviet Union for leadership in space, but Kennedy and Johnson also saw the National Aeronautics and Space Administration as an additional opportunity to achieve progress on Earth—"progress for all." (Even then the ISS motto "Off the Earth, For the Earth" would have been appropriate.) As a 2014 *Smithsonian Air & Space* magazine article titled "How NASA Joined the Civil Rights Revolution" so aptly put it, "History forced President John F. Kennedy to commit the country to explore space at exactly the same time it forced him to confront the movement for civil rights."[2]

One of Kennedy's integration strategies was to focus on federal employment. NASA began actively hiring African-American scientists and engineers from around the country in the late 1950s to support the space program. These hirings to expand the diversity of the NASA workforce also included the talented African-American women mathematicians who became known as the "computers" for the space program: They performed some of the most important calculations for getting humans in spaceships safely to and from space. Now recognized as "Hidden Figures," mathematicians like Katherine Johnson and Margaret Hamilton were indeed "hidden," but even then it was no secret at NASA that they were critical to making it all happen. When Margot Lee Shetterly's best-selling book by that title came out in 2016, followed by a blockbuster film the same year, these women finally were recognized for their significant contributions.[3] In 2020 the NASA headquarters building in Washington, DC, was renamed after Mary Jackson, the first African-American female engineer at NASA.

In the 1960s, as the race to the Moon accelerated, minority hiring at NASA also accelerated through internship and hiring programs established in partnership with the historically black colleges. As NASA prepared to take humans farther off the planet, it was launching equal employment opportunities for deserving, yet underrepresented, humans here on the planet.

NASA was also an early proponent of increased diversity within the astronaut corps. While in 1959 the first group of NASA astronauts, known as the Mercury 7, was made up entirely of white, male, military test pilots, in 1962 NASA opened up the second class to civilian test pilots too. This might not seem like a big deal in the grand scheme of overall diversity, but it definitely was a big deal to pilots. In 1967, Ed Dwight, an Air Force test pilot, was the first African-American to be put forward by the Air Force for astronaut selection. While a progressive choice in light of the broader social dynamics of the time, he unfortunately never flew as a NASA astronaut. The seventh class of astronauts, selected by NASA in 1969, was a military group that did not increase the racial or gender diversity of the astronaut corps, but the next class, which wasn't selected until nine years later, in 1978, sure did.

The Space Shuttle program opened up the opportunity to be an astronaut to a much broader demographic. A new kind of spacecraft with a new mission meant that NASA would need a new kind of astronaut. For the work that would be performed on the Space Shuttle, NASA needed to expand the expertise beyond just pilot astronauts to a new designation they called "mission specialist," which included scientists and engineers. NASA also wanted the astronaut corps to better reflect our population with respect to race and gender, but they struggled on their own to get the word out to candidates from this more diverse population and encourage them to apply. One of my favorite stories about the recruitment for this first class of Space Shuttle astronauts involved leveraging the power of sci-fi to bring sci-fact to reality—and it involved the best sci-fi series of all—*Star Trek*.

Star Trek, which first aired as a TV series in 1966, has from its beginning been an inspiration for the evolution of technology, but through the intent and forward thinking of its creator Gene Roddenberry, it also challenged cultural norms with its multicultural and multiracial crew—a Japanese helmsman, a Russian navigator, an African-American female communications officer, and a human-Vulcan first officer—and plenty of encounters with other intelligent alien life from across the universe. As a young child, some of my earliest memories of TV are of watching *Star Trek* with my family. I watched with excitement about the crew's "missions," but more importantly, I didn't question the types of people that were on the crew. Because of *Star Trek*, I grew up thinking that crews of space missions looked like the ones on the deck of the *Starship Enterprise*. I'm thankful that the crews I flew with looked like this too.

After the original *Star Trek* TV series went off the air in 1969, Nichelle Nichols, the actress who played Lieutenant Uhura (the

African-American female communications officer), continued her acting and music career, but through appearances at *Star Trek* conventions she was introduced to NASA scientists and senior management. She became a voice for women and minorities in the space program. She believed that our space program is for everyone, and she posed the question: "Where are my people?" In 1977, NASA hired her through her company Women in Motion, Inc. (which she had originally formed to use music as an educational teaching tool) to help the space agency overcome the challenge it was having with astronaut selection for the Space Shuttle program. Already eight months into its recruitment process, NASA had received only 1,500 applications, with fewer than 100 from women and only 35 from minorities. When Nichelle finished her intensive four-month recruitment campaign across the nation, NASA had over 8,000 applications—1,649 from women and over 1,000 from minorities.

In Nichelle's NASA recruitment video, she said, "If you qualify and would like to be an astronaut, now is your time to apply. This is your NASA. NASA is a space agency embarked on a mission to improve the quality of life on Earth right now."

NASA was so impressed with Nichelle's recruitment results that in 1978 it increased the selection for the first Space Shuttle astronaut class roster from twenty-five to thirty-five, to include the first three African-American men, the first Asian-American man, and the first six women. Nichelle Nichols continued to support NASA recruitment and in 1984 was presented with the prestigious NASA Public Service Award for her many efforts toward an integrated US space program.

In March 2021, there were seven international crew members on board the ISS—two women, one African-American man, and four white men, including one Japanese and two Russians. As of January 2021, the astronaut office has forty-six astronauts, 35 percent of whom are female, which is well above the mean when you consider that it's nearly double the typical 20 percent enrollment of female students in collegiate engineering programs and that, as of 2018, only 13 percent of the entire US engineering workforce were women.[4]

If you looked at a 1960s photo of the front room of the launch control center at the Kennedy Space Center or the back room of mission control in Houston for the Apollo missions, you'd see a sea of young white men wearing button-down, short-sleeved shirts with pocket protectors and only one woman among them in each location: JoAnn Morgan in Florida and Frances "Poppy" Northcutt in Texas. Today, not only are both the launch and mission control centers and all of NASA human spaceflight *run* by women, but within those control centers and across NASA there is a wonderfully diverse mix of humanity at work.

At the time of this writing, NASA is planning to send more astronauts on a mission to the Moon as early as 2024. At least one of those astronauts will be a woman.

The US space program began as a literal race to the Moon against the Soviets. While there were many good reasons for this race, I'm pleased that today we recognize the value of international partnership and cooperation and don't focus so much on competition. As is human nature, competition will always exist, but the space programs around the world that were once competing in space against one another have demonstrated through the ISS program that when we practice peaceful competition and cooperation instead, we all experience the benefits.

In 1975, as part of the historic Apollo Soyuz Test Program (ASTP), NASA astronaut Tom Stafford and Soviet cosmonaut Alexey Leonov shook hands across the open hatch of their US Apollo and Soviet Soyuz spacecraft. The success of ASTP and the handshake between these two spacecraft commanders signaled to many that the space race was over and demonstrated not only that we could connect two spacecraft from two countries together in space, but also that a meaningful connection between the people of those countries could be made.

It was also the beginning of a lifelong friendship for Tom and Alexey.

In a 2013 article by space journalist Dr. Elizabeth Howell that celebrated the Apollo-Soyuz mission, she observes that even while the ASTP was still in the planning stages, talks were going on behind the scenes about cooperation between the US and Soviet space programs on future space exploration activities. As Dr. Howell so eloquently puts it, "Beyond the technology, it [ASTP] would also be a demonstration to the world of peace and shared purpose."[5] The ASTP mission laid the groundwork for international cooperation in space with initiatives such as the Shuttle-Mir Program in the 1990s, the ongoing International Space Station program, and today's space exploration plans to travel back to the Moon and on to Mars.

Astronauts from different nations began flying on the Space Shuttle in the early 1980s, and NASA astronauts started flying on the Russian Soyuz spacecraft in the 1990s. After the Space Shuttle program was retired in 2011, US astronauts continued to fly with our Russian partners on the Soyuz to and from the ISS. The final planned flight of a US astronaut on the Russian Soyuz to the ISS took place in April 2021. The second successful flight of a SpaceX-crewed Dragon spacecraft to the ISS took place in November 2020, and the international crew on this second flight included an astronaut representing the Japanese Space Agency (JAXA). This mix of international crews on both Russian and US spacecraft is planned to continue. I would argue that any tensions between these nations are tempered by our cooperation in space.

A couple of weeks into my first stay on the ISS, I was surprised by a call from fellow astronaut Brent Jett, who at the time was serving as NASA's director of flight crew operations. He'd radioed to tell me of my assignment to a second space flight. Along with two of my Bug classmates, Mike Barratt and Tim Kopra—who also were my crewmates at the time onboard the ISS—and three more Bug classmates on the ground—Eric Boe, Al Drew, and Steve Bowen—we would join our commander Steve Lindsey (a Flying Escargot) on the STS133 mission of the Space Shuttle *Discovery*, which had been designated as the final flight of the Space Shuttle program. Never before had astronauts been assigned to their next mission while they were still in space on a current mission. I would be rolling directly into training for my second flight immediately after my return from three months in space on my first flight. Happy as I was, I was both excited for and dreading the phone call I had to make from space to my husband with this news.

In February 2011, five Bugs and one Flying Escargot (now an honorary Bug) flew the STS133 mission, spending two weeks at the ISS performing assembly, maintenance, and science tasks. To look at the six of us, you wouldn't have seen anything particularly outstanding about any member of the crew (except our freakishly good looks, *har-har*). Without any other knowledge of who we were or the diverse experiences we'd had, you might have seen only one white woman, one African-American man, and four white men. Beneath the surface, though, you would have found a world-renowned flight surgeon, an artist, a missionary, five engineers, a chef, a special forces officer, three fighter and test pilots, five parents, a sailor, six scuba divers, a submariner, and a couple of carpenters. That's a lot of roles for six people, and it wouldn't have begun to capture the depth of diversity that was present. In short, you would have found a group of people with diverse experience as well as expertise across all necessary astronaut skills, like piloting a Space Shuttle, flying the robotic arm, spacewalking, performing Space Shuttle and ISS systems operation and maintenance, and conducting the science of the mission. And most importantly, you would have found a mix of people who believed in the overarching mission of spaceflight: that it's all about improving life on Earth.

Although NASA selected us for our technical skills, on the public side they believed in our ability to thoughtfully communicate this mission to the general public and the international community, especially during the historic transition of retiring the Space Shuttle program.

There is always more to each of us than what you can see on the surface. There is always more to the life that surrounds us than you can see from a distance. When we look at Earth from space, we see one big, beautiful planet. If you were in space on the ISS, what you would see through the window with your naked eye is a planet that looks alive. All the blues, greens, browns, whites, and the myriad of other colors you know Earth to be are presented in crystal-clear high definition. You would see movement—the rotation of the planet itself; the speed at which we're traveling around it; the white clouds swirling around; the shadows of the clouds that seem to move along with them; lightning strikes that chase one another and wrap around the planet; wavy translucent green curtains of aurora floating above; the rainbow light in the shift from night to day; the city lights "coming on" as the spaceship moves over the Earth at night and turning off as it passes into day; and the changing glint of light on the water depending on the angle of the Sun. Yet it's impossible to see any of our planet's life-forms from space unless we zoom in. We can't begin to understand what's happening on our planet unless we look closer still. Zooming in enables us to see and understand the details. It gives us the means to appreciate even the smallest life-forms that live among us. It provides a way for us to understand the significance of even the tiniest forms of life for all us Earthlings.

The astronauts of our NASA class of Bugs have received many accolades for the critical and significant roles we've played across our many space missions, but *actual* bugs—the insect, worm, and spider kind—play an even more critical, significant, and largely unrecognized role in the survival of all life here on Earth. Bugs are a massive and unimaginably (at least to me) diverse group, and for centuries they have gotten a bad rap as creepy, crawly, gross pests. Most human actions toward bugs have been attempts to eradicate them from our farms, gardens, and homes by mass, indiscriminate extermination.

I think a lot of people have some sense of the fact that the Earth houses a great variety of bugs like bees and other insects, as well as worms and spiders, but I was shocked by just how many are known to be living on Earth—there are some 900,000 different kinds. For some perspective, consider this: there are more species of ladybugs than mammals, more species of ants than birds, and more varieties of weevils than fish. The true number of living species of insects can only be estimated from scientific studies. Most authorities agree that there are more insect species that have *not* been described—meaning named by science—than the ones that have been named already.[6] This means that there are at least two million species of insects at a minimum, but scientists estimate there are over five million. That doesn't begin to approach the total *number* of insects. The total population of insects, at any given time, is estimated at some ten quintillion (10,000,000,000,000,000,000) individual insects.[7] I don't know about you, but I'd never even heard of the number "quintillion" before I looked this up. As far as I'm concerned, a quintillion could be a gazillion, and this count doesn't even include the over one million species of worms and over forty-six thousand (likely up to one million) species of spiders that also are recognized as "bugs."

Along with this ginormously diverse group comes a wide variety of "expertise" in bugs, which perform a lot of amazing functions for us all. Like the vital role the astronauts play in ensuring crew survival on the Space Shuttle and the ISS, bugs are vital to the survival of all life here on Earth. So much of the beauty and bounty of nature that we take for granted, like flowers, fruits, and vegetables, and the not so beautiful side of the process of decay and recycling nutrients back into the soil are driven by the vital role that bugs play in the Earth's ecosystem.

With such massive, mind-boggling numbers of bugs on the planet, it may seem impossible to imagine that we humans could have any kind of significant impact on their overall population or their ability to survive. Unfortunately, we do.

A 2018 global study on insect populations warned that over 40 percent of all insect species are in decline. The potential detrimental impacts of this decline on humanity outlined in this study led to the idea of a looming "Insect Apocalypse"—aka "insectageddon"—and it continues to be a topic of discussion and exploration among scientists.[8] While some have argued that these estimates exaggerate the threat to insects and our planet, in February 2020 a team led by Dr. Pedro Cardoso reported in *Biological Conservation* that "the current (insect) extinction crisis is deeply worrisome. Yet, what we know is only the tip of the iceberg."[9]

Human activity—including mass extermination with pesticides, pollution, and habitat destruction and encroachment into previously untouched lands—has led to a decline in bug populations. This should get everyone's attention because a world without insects is a world without humans.

"Insect declines lead to the loss of essential, irreplaceable services to humanity. Human activity is responsible for almost all current insect population declines and extinctions," Dr. Cardoso says. "The fates of humans and insects are intertwined.... Solutions are now available—we must act upon them."[10]

Myriad solutions have been proposed from too many sources to name here. These solutions range from bigger nature reserves and a crackdown on harmful pesticides to individual actions such as leaving dead wood in gardens and not mowing the lawn. Scientists also urge that invertebrates must no longer be neglected by conservation efforts, which tend to focus on mammals and birds.

As to the missing damselflies that were such a vivid part of my childhood days at the airpark, I am always on the lookout for them now. In my quest to find out about their current status, I made a fortuitous connection online with Dr. Dennis Paulson, a dragonfly and damselfly expert. Dr. Paulson was extremely generous with his time as he answered my out-of-the-blue query: "Where did all the damselflies go?"

He shared some insight into how other creatures have been studied. For example, for over one hundred years, since 1900, the National Audubon Society has been conducting the Christmas Bird Count in the United States—an annual, quantitative approach to understanding bird populations. The North American Butterfly Association has been counting butterflies since 1993 through the Fourth of July Butterfly Count. However, Dr. Paulson told me, "Insects have scarcely been censused at all, so we don't have much in the way of baselines to judge how things are going over time. There are fortunately a few studies with recent commentaries on insect decline that are well quantified, but surprisingly little from North America."

Dr. Paulson has lived in Seattle since 1967. Prior to that, he lived for fifteen years in Florida, where he earned his PhD in zoology at the University of Miami. "I wrote my dissertation on the dragonflies of the region," he told me. "Unfortunately, it got so long (at 603 pages, it was the longest ever written at the U. of Miami at the time) that my advisor asked me not to write up the damselflies, and I complied. But I have been forever sorry about that, as that would have been worthwhile, and I just never got around to it."

Dr. Paulson has maintained a leadership role in the science of insects, including coauthoring some of the most current and respected papers on Odonata, the order of flying insects that includes dragonflies and damselflies. His work is referenced in some of the leading scientific papers addressing the insect crisis, he has authored several field guides for dragonflies and damselflies, and he was the director of the Slater Museum of Natural History in Tacoma from 1990 to 2004.

While my observations are not the same as a quantitative study, Dr. Paulson affirmed that it's not my imagination that I don't see damselflies in the same swarming numbers anymore. He shared a similar story about his experiences looking for giant water bugs in Florida. "I used to haunt well-lit shopping centers and other public areas around Miami, and they were alive with insects of all kinds," he told me. "Giant water bugs used to be in numbers such that I had to watch for them flying around and try to avoid one landing on me and possibly biting. Now, in similarly brightly lit settings at night in natural habitats that aren't that much changed (e.g., around the edge of the Everglades or Big Cypress Swamp), there are no giant water bugs and shockingly few other insects."

This is the same experience I've had with the damselflies (except for the biting part). Dr. Paulson went on to point out that "dragonflies and damselflies in general are still common in comparison with other insect groups."

I was happy to hear this about the damselflies. At the same time, I was sad—sad and angry and frightened—because I think the sparseness of damselflies today speaks to the more serious issue with insect populations overall. If I can remember having seen a damselfly only one time since the 1970s—in 2019, while having dinner outdoors at a restaurant in Sarasota—and their populations are some of the *least* declined, then I'd imagine there are a lot of other insects unknown to me that I'll never have a chance to see.

Regardless of whether scientists agree on the actual numbers of insect decline, they all seem to be aware that the loss of even a small percentage of insects could be disproportionately consequential. Bugs sit at the base of the food web; if they go down, so will many birds, bats, spiders, and other predators. Bugs aerate soils, pollinate plants, and remove dung and cadavers; if they disappear, entire landscapes will change. Given these risks, "do we wait to have definitive evidence that species are disappearing before we do something?" asks Dr. May Berenbaum, head of entomology at the University of Illinois and the 2014 winner of the National Medal of Science. "There are so many connections [between bugs and all life on Earth] that we haven't even begun to appreciate. We usually find out when they go missing, which is not the best way to find out," she said. "Don't dismiss an animal because it's tiny. It could be playing an oversized role in the life of some other organism."[11]

As I have come to understand how important it is for us to recognize our planet as our only home, one thing that Dr. Cardoso said in the 2020 report has stood out the most: the global economy, he wrote, "is fully dependent on ecosystems, as natural resources are the basis of everything. Insects are the major part of ecosystems, and without them nothing works. It is no coincidence that ecology and economy have the same origin 'eco,' meaning 'home.' There is no economy without ecology."[12]

Bugs aren't capable of discerning which plants have been laced with harmful pesticides; neither can they relocate across long distances in response to widespread land clearing for agriculture, and they definitely don't have the ability to alter human activity or curtail rising temperatures. It's up to us to change our ways or risk losing these tiny yet essential contributors to life.

Like so many environmental issues, the task of saving bugs and thus saving ourselves can seem beyond the power of any one individual, but there is hope. Dr. Helen Spafford, a University of Hawaii entomologist and coauthor of the Entomological Society of America's 2017 position statement on endangered insect species, says of insects, "They're the unsung heroes of most ecosystems." Scott Black, executive director of the Xerces Society, points out that "the neat thing about insects is, anybody can help them. If you have a little yard, if you're a farmer, if you're a natural area manager, if you work at a department of transportation, you can work to manage plants for pollinators. We can do this across the landscape and we need to."[13]

I found these words inspirational for my own life, and I believe you can take them to heart in your life as well. When I improved the small yard around my home, I deliberately cultivated a garden where insects could thrive. I receive so much joy at the sight of butterflies and bees flying around the garden and worms in the dirt!

There are many ways to make room for bugs, and not only around your own home. Consider starting or contributing to a community garden or encouraging others to plant bee- and butterfly-friendly flowers and plants near where you work or live. Insect-friendly gardens, Dr. Cardoso said, can supplement large-scale solutions and help halt the decline. "When lots of people implement these small solutions, it can make a big difference to many insect populations. Even a couple of gardens could be a big thing for a species."

Just as our actions can seem small and arbitrary and unworthy, bugs are often considered small, arbitrary, and unworthy and often thought of as having little to do with human everyday life. But in reality, both little bugs and small solutions really do make a big difference.

It's clear that bugs are essential to the survival of life here on Earth. As we continue to inhabit and explore space off the planet for longer periods and plan for human expeditions deeper into space, scientists have been studying how we might apply the benefits that bugs provide us here on Earth to establishing life on the Moon and Mars. Many studies have been (or are being) performed on the ISS regarding the benefit of bugs. Two examples of bug studies in space that are particularly interesting to me are the ones that involve honey bees and microscopic worms; each of these studies could help improve life on Earth as well as enable exploration farther off our planet.

The honey bee studies are a vital link in explorations of how we might one day bring the pollinator with us to the Moon, to Mars, and beyond to support off-Earth agriculture and possibly have delicious, healthy honey in space. Initial studies of bees in space show that, like most animals, they don't do very well in sterile, isolated environments. They are dependent on the nature around them to thrive. So, while hopeful that bees could eventually thrive in a spacecraft environment, scientists also are investigating the potential of man-made micro-drones to mimic bees as pollinators in places where bees themselves might not thrive. It terrifies me to think, though, that if we screw up nature badly enough, one day we may need to use micro-drone bees created for off-world pollination to perform pollination here on Earth.

Another experiment performed on the ISS may deliver more immediate benefits by leading to a way to protect crops without destroying beneficial insect neighbors, both on Earth and in space. Pheronym, Inc., is a Florida-based bio-ag-tech company and recipient of the Biomimicry Institute's 2020 Ray of Hope Prize, which celebrates and accelerates nature-inspired solutions to the world's environmental and sustainability challenges. Biomimicry is the design and production of materials, structures, and systems that are modeled after biological entities and processes. The Biomimicry Institute defines it as a practice that learns from and mimics the strategies found in nature to solve human design challenges (and to inspire hope along the way).

Pheronym has already developed eco-friendly pest control methods here on Earth via biomimicry by utilizing microscopic roundworms (nematodes) that have naturally occurring bacteria in their gut that kill crop-harming insects, but that are also nontoxic and pollinator-friendly and are therefore safe for beneficial insects like honeybees. Once the nematodes kill and eat their prey, they release a pheromone (a secreted or excreted chemical factor that triggers a social response in members of the same species) as a signal to other nematodes that they need to find and infect another insect pest. Pheronym has developed a way to use this pheromone signaling to accelerate nematodes' pursuit of their prey and increase their effective role as pest control.

These worms have now been studied on the ISS, where the focus was on ecofriendly pest control for off-Earth agriculture and the microgravity environment was utilized to demonstrate that the worms can effectively function there. Dr. Fatma Kaplan, CEO of Pheronym, said that the ISS research "gives us valuable insight on how to keep beneficial nematodes alive and viable for agriculture on other planets."[14]

It's pretty impressive to think that a microscopic nematode can help us grow food in a nontoxic and sustainable way both on and off our planet. I am in awe of the role that each of us play as diverse creatures on this planet. I am in awe of the planet itself and how perfectly calibrated it is to sustain us (if we'd only respect it). I am in awe of the way this all makes me wonder about my own place on Earth, and how I can better appreciate, better respect, better "see" the planet itself and all the creatures I share it with. Bottom line: to appreciate the awe and wonder of all living things, we all need to zoom in and look closer.

You don't have to travel to space to appreciate the awe and wonder that surrounds us every day here on Earth. Astronauts are always trying to find ways to relate the impact of the experience they had viewing the Earth from space with experiences here on Earth. I've discovered, though, that the best way to appreciate that awe and wonder here on the planet is to look closer, to see the things that we treat as familiar in a whole new way, and to open up our eyes and hearts and minds to the details in all of their simplicity and complexity.

As I searched for someone to showcase who is creatively sharing a closer look at the awe and wonder of what surrounds us right here on our planet, I was introduced to the photographer and author David Liittschwager and his 2012 project, "A World in One Cubic Foot: Portraits of Biodiversity." His project began with the question: how much life can be found in a small piece of the world? As soon as I researched some of his photography, I discovered that he was the one who took an iconic nature photo that had been important for my own understanding of the plastic problem on our planet. This image showed the insides of the stomach of a baby albatross that was discovered dead on the Kure Atoll in the far reaches of the Pacific Ocean. The bird's stomach was full of hundreds of tiny pieces of plastic. David photographed the bird with its open stomach overflowing with plastic, and the photographer Susan Middleton took an accompanying photo of all the plastic neatly arranged in a circle against a stark white backdrop. This pair of images is shocking and sad to see. It's clear that this poor bird didn't stand a chance. These images are on display in many places, including from October 2019 to September 2021 at the American Visionary Art Museum in Baltimore, Maryland, as part of an exhibit called "The Secret Life of Earth: Alive! Awake! (and Possibly Really Angry!)" I love the title of the exhibit—especially that last bit.

I had the good fortune to interview David, and I was pleased to meet a humble and thoughtful human being who demonstrated a deep sense of awe and wonder toward the world around him. As I did in all the interviews for this book, I asked him if he'd had an Earthrise moment. I loved hearing his immediate response. He said, "Well, I hope I have them kind of regularly."

His words made me realize that we should always be appreciative of the revelation that can come from experiencing the world around us. David's Earthrise moments, he continued, come "from geology and biology. I mean, I think the understanding of the way the world's made that one gets from science is—there's no lack of spirituality or things that generate a sense of awe. I think knowing how the world works, and seeing how beautiful it is, is awe-inspiring."

He shared with me a little about his time on Kure Atoll. "I've had the great privilege to be sent to go see things that I didn't have any knowledge of ahead of time," he said. "At one point I was standing on the northernmost edge of Kure Atoll, which is the northernmost of the Hawaiian Island chain. It would be, in some ways of looking at it, the farthest place away you could go on Earth."

Designated as part of the Papahānaumokuākea Marine National Monument, Kure Atoll is one of the largest preserved and protected marine areas in the world today. The effort to bring the atoll back to its natural state took more than just drawing a line around it or erecting a fence. "In order to support and protect the life that's historically been there, and restore it to a magnificent jewel, we needed to stop this not native, nasty weed (*verbesina encelioides*) that found its way onto the island as seeds on the tires of U.S. military trucks carrying supplies," David said. He saw how the invasive weed had devastated the habitat, in particular the habitat for ground nesting birds like the albatross, to the point where the birds couldn't even fledge because the overgrowth of weed deprived them of the room necessary to grow and take flight. Now a member of the board of directors for the Kure Atoll Conservancy, David can attest to the work that's been done by a very small group of people to restore the island almost back to its original condition.

"They've succeeded. The albatrosses are not being pushed out. I've seen what the work of one small group can do. I know for now that the albatross have the space to learn how to fly that they didn't have a decade ago. That's a really cool thing."

David explained to me why it's important to not just photograph life but to make a portrait by zooming in and getting the details—like the detail in David's photograph of the belly of the baby albatross. "People sometimes complain about my pictures that the habitat's not included, or that there's not some interaction with another creature," he said. "Those are great pictures too, but I wanna see the creature, I want to make a portrait of the creature. If it's a large mammal or bird or some creature that I can capture a gaze that could be conceived as some sort of mutual regard—I like that a lot." He said that you can find this mutual regard with most creatures, "even with ones who are tiny or so foreign to us that we can't imagine it accurately."

Take the octopus. "They're certainly capable of mutual regard. When I've photographed them, they're very aware of everything that's going on around them, and they respond to visual direction. The thing is, they have a little brain, most of them are small, and so their brain isn't even the size of a pea. But it's really not about their brain. When they touch something, they can taste it. Without the signal traveling all the way to the brain and back, the texture and color of their skin changes to meet their background as they pass over it or touch it. That's something that's super cool and something we don't have any experience with. I mean, that's just so foreign."

"Some would say alien," I quipped.

I was joking, yet David's words describing his encounter of "mutual regard" with an octopus did remind me of my moment of mutual regard with the damselfly that sat on my finger at the Clearwater Airpark.

David shared a thought he'd had about the social distancing we were all experiencing at the time because of the COVID-19 pandemic. "We have to be willing to 'look.' You can be really kind from six feet away, and that's the thing to do right now. It's amazing how much you can still notice from a distance if you take the time to really look and see. We should all be paying better attention—if you ask me if I have anything that I regret, it's that I haven't paid attention enough."

Some might argue with David's self-assessment of how much attention he pays to life around him. The wonderful partnership between David as a photographer and the scientists he brought together through the "One Cubic Foot" project has integrated the art and science in a beautiful, visual presentation of the unseen and abundant life that surrounds us.

Sitting around the dinner table with his longtime girlfriend Susan, David told me, "I think she's the one who first said 'one cubic foot.' Can't really pin it down, but it's something that fits in your lap or you can put your arms around." A sample size for any study needs to be manageable, but it must also accurately represent the environment being studied. David chuckled when he told me, "You'd think that one cubic foot is a manageable sample size, but my dirty little secret is, I never finished [cataloging] a single one of them. If you just increase the resolution a little bit or spend a little more time at it, you're always gonna find more."

The implementation of the project was pretty straightforward: David took a metal frame that measured one foot by one foot by one foot (the one David built was green), placed it in a location of interest, and then captured individual portraits of all the life that it contained and that passed through it over the course of one day and one night—anything visible to the naked eye, no matter how small. Selection of the location was guided by scientists.

David and his scientist partners traveled to different habitats across the planet: the deciduous forest in the Hallett Nature Sanctuary in New York City's Central Park; mountain fynbos in Table Mountain National Park in South Africa; a cloud forest in the Monteverde Cloud Forest Reserve in Costa Rica; a coral reef at Temae Reef in Moorea, French Polynesia; a freshwater river at the Duck River, Lillard's Mill, Tennessee; and a saltwater bay under the Golden Gate

Bridge in San Francisco. The samples from each of these locations offer endless opportunities for study by botanists and -ologists of all kinds; for the layperson like me, they offer another opportunity to witness the abundance of life that surrounds us and to develop a better understanding of the diversity that's necessary to ensure survival for us all. (I was thrilled to see that David had captured images of some damselflies during his one cubic foot survey of the freshwater river in Tennessee.)

David marveled at an idea that one scientist shared with him about the placement of the cube in the South Africa location. "He described that while seated on my butt, on the ground, I would be able to reach out and touch fifty species of plants. Move the frame a little to the right or left and you get a different selection of plants and animals." Each of these South African plants and animals plays a role in the lives of the other plants and animals it shares that cubic foot with.

"The location chosen in the middle of New York City's Central Park was an area that had a fence around it since 1939, and it just hadn't been stepped on much. It was sort of this second-growth eastern deciduous forest. It was wonderful to see it thriving, but at the same time it's like there was one group of insects that would normally be quite numerous that were completely absent. That was the night-flying insects, because they would be drawn off every night because of all the city lights. It's really wild."

This reminded me of the delicate balance of life on our planet. Not only are we dependent on one another, but if our planet was located just a little bit closer to the Sun or a little bit farther away from it, life as we know it could not exist on Earth.

David has been influenced significantly by the work of E. O. Wilson. (I'd say we all should be influenced by his work.) Wilson is recognized as one of the most important biological theorists since Darwin. He has transformed his field of research—the behavior of ants—and applied his scientific perspective and experience to illuminate the human circumstance, including human origins, human nature, and human interactions. Wilson also has been a pioneer in spearheading efforts to preserve and protect the biodiversity of the planet through the Half Earth Project, which is a big idea in that it could save not only nature but humans along with it. Through his E. O. Wilson Biodiversity Foundation, he is on a mission to promote worldwide understanding of the importance of biodiversity and the preservation of our biological heritage.[15]

David shared with me some of E. O. Wilson's ideas that he thought I might find particularly interesting.

"It may seem that the whole icky lot of them [bugs], and the miniature realms they inhabit, are unrelated to human concerns. But scientists have found the exact opposite to be true. Together with the bacteria and other invisible microorganisms swimming and settling around the mineral grains of the soil, the ground dwellers are the heart of life on Earth. And Earth is the only planet we know that has a biosphere. This thin, membranous layer of life is our only home. It alone is able to maintain the exact environment we ourselves need to stay alive. Most of the organisms of the biosphere, and the vast number of its species, can be found at the surface or just below it. Through their bodies pass the cycles of chemical reactions upon which all of life depends. With precision exceeding anything our technology can match. In time we will come fully to appreciate the magnificent little ecosystems that have fallen under our stewardship."

As I think back to the year 2000, when our astronaut class was named in jest for what turned out to be a comical fear that the world would collapse due to the fictitious "millennium bug," I find it both sad and ironic that the looming disappearance of essential bugs, caused by a general lack of appreciation for the diverse functions that they perform, and continued ignorance of the interconnectivity and interdependence of all life on Earth might actually lead to our collective extinction. We still have the opportunity to reverse this course, but first we must acknowledge the significance of all life and recognize that often what seems to be too small to be significant can have the greatest impact.

From the flight deck of the Space Shuttle *Discovery*, remotely guiding my crewmates Al and Steve through their spacewalk activities outside the International Space Station.

NASA

CHAPTER 5

GO SLOW TO GO FAST

When asked before making the very first crewed Space Shuttle flight if he was "worried," John Young, one of the best astronauts ever and the commander of that mission, replied, "Anyone who sits on top of the largest hydrogen-oxygen fueled system in the world, knowing they're going to light the bottom, and doesn't get a little worried, does not fully understand the situation." This from the first man to fly in space six times, including two Gemini missions

(Gemini III and X), two Apollo missions (Apollo 10 and 16, where he both walked and drove on the Moon), and two Space Shuttle missions (STS-1 and 9).

There's nothing slow about launching into space except all the preparation that goes into it. Most astronaut training focuses on preparing to respond to everything that might kill us. A lot of things can go wrong when you're launching to space on top of seven million pounds of exploding thrust, then working on a space station with only a thin metal hull between you and the deadly vacuum of space. Three things that can go wrong and are classified as emergencies are fire, toxic atmosphere (when something toxic like ammonia gets inside your space station), and rapid depressurization (when the air in your space station escapes out to the vacuum of space). Each situation is life-threatening and requires immediate action.

About a week into my first mission, I was awakened in the middle of the night by an abrasive, high-pitched emergency alarm, which sounds a lot like a home smoke detector, only louder. Being awakened by an alarm in the middle of the night on a space station got my adrenaline pumping, but I didn't feel panic. My training kicked in, and I was ready to go to work and do my part to handle the possible emergency. Floating out of my crew compartment, I was happy to see that my crewmates all were doing the same. Everyone responded as we'd been trained, taking the required actions with deliberate calm. Each of us made sure that all of the crew were accounted for. We looked out for one another and took the immediate responses of silencing the alarm, communicating with mission control, and working through the checklist together to troubleshoot and resolve the problem.

That night we were responding to a depressurization alarm. Thankfully, there was no hole in our station. Once we were able to verify that our precious air wasn't leaking out to space, we each floated back to our crew compartment to go back to sleep (though maybe a little less soundly than before).

The most impressive thing to me, and what gave me the most pride and confidence, was not our technical ability to manage the problem, but how well we did it as one unified crew. Emergency alarms sounded periodically throughout the mission, and each time we responded as one crew to solve the problem for the whole station. We never thought about which part of the space station might be impacted. No one hoped that it was in the Russian part of the station and not the American part, or vice versa. We knew that if we failed to resolve any emergency in any part of the space station, it could mean the end for all of us.

In the astronaut world, we have a mantra: "slower's faster." What I've learned to be true in any situation, especially in an emergency, is that it's less about how quickly you act and more about understanding when you need to act and how precisely you execute the necessary steps. Precision only comes through calm, which comes through deliberate practice. In short, the philosophy of "slower's faster" is about being prepared and getting things right the first time.

The protocol NASA uses to implement the "slower's faster" approach in a spacecraft emergency is to have each crew member memorize and diligently and deliberately execute the immediately necessary steps—a procedure called the "boldface" in the survival checklist. Once we've executed the boldface, the spacecraft is in a safe configuration and we have the opportunity to consult with mission control on the ground and carry out more extensive procedures to figure out what happened, get back to normal operations, and try to prevent the emergency from happening again.

Without going into the details of the protocol for how we handle each life-threatening situation we might encounter, there are two overarching themes to every boldface checklist for managing every potential problem:

1. *Crew survival.* We define crew survival as "crew" survival, not "part of the crew" survival. So the first step is accounting for all your crewmates. Never leave a man (or woman, thank you) behind.
2. *No freaking out.* Not that a fire on the space station wouldn't get the adrenaline pumping, but we have to put the adrenaline to good use. We train so we can apply that energy boost to executing the emergency response with quiet, calm confidence, and without hesitation.

Caesar Augustus, the first emperor of Rome, who reigned from 27 BC until his death in AD 14, had a favorite saying, *Festina Lente*, which means "Make haste slowly."[1] He was so fond of it that Roman coins were minted with various images that symbolized the saying—a crab and butterfly, a rabbit in a snail shell, a dolphin wrapped around an anchor (which seems to have been the one that stuck). My favorite version of the coin depicts a tortoise with a sail on its back. Augustus often communicated this sentiment in speeches, in which he was known to say, "That which has been done well has been done quickly enough," meaning that activities should be performed with a proper balance of urgency and diligence.

The "make haste slowly" mantra has been primarily associated with innovations in warfare. The Navy SEAL version is "Slow Is Smooth. Smooth Is Fast." Underlying this is the idea that SEALs do not need to rush because their movements are strategic. You better believe that Navy SEALs train extensively before going into combat. To watch them in action is to see a group proceeding as a coordinated unit with deliberate calm and bursts of movement only as necessary.

If you were to watch an ISS crew responding to an emergency, I think you would see actions similar to those of the SEALs. Refined and deliberate, simple and elegant, our movements would appear more like a choreographed dance than a haphazard, disorganized, or panic-fueled reaction. And while we can't predict the precise time or way in which any emergency will present itself, we can be confident that we're prepared to respond in the most effective way possible based on what we know in advance.

At NASA, we know that preparing for what we consider known or potential threats with slow and careful forethought results in safer and more successful responses than hurrying unprepared into action. This is why we train the way we do, and why we do our best to be proactive in how we prepare to react. We have classified everything we think we know can go wrong in space into categories like "caution," "warning," and "emergency," depending on how quickly action is needed and the level of threat to the crew and station.

Our training also utilizes a crawl-walk-run approach to responding to emergency situations. By the time we complete our training and are ready to fly, we understand who is doing what and when. We've worked together through the boldface checklist in a deliberate fashion to the point where we have an almost unconscious competence because we know our tasks, we know the required actions, we know our crew, and we are working together smoothly, urgently, and with purpose. As we refine our training, our performance continues to improve. We develop the ability to respond at reduced speeds that make us faster when we need to go full speed. While there is such a thing as too slow, we gain the understanding that full speed is often slower than we think. There is a difference between going full speed and being rushed.

As many business leaders have professed, this approach applies to a multitude of other situations; I would argue that it applies to pretty much every aspect of life where the outcome is critical. Whether in a military operation, on a spaceflight, or in any other scenario, high-performing teams (crews) know they need to be able to rely on one another during the worst of times. Each crewmate, using their unique skills, works to protect the others. The overlapping skill sets and high performance of each crew member knit together to form a unit that is stronger than its individual parts. In a crisis, this enables us to focus on crew survival without freaking out.

The best crew members also develop what we call good "situational awareness"—the ability to mentally process and understand multiple things happening in the environment around them and to make decisions and/or take action in a relaxed and purposeful way. Our training allows us to relax and perform the steps with efficient precision. When we're relaxed and calm and need to move fast—even when we think the air might be spewing out of the space station or something's on fire—we find ourselves assessing all the information available to us and moving as necessary, without thinking about the time involved. This is why I like the alternative way that Augustus defined "slower's faster": "That which has been done well has been done quickly enough."

In 1998, I interviewed for that year's astronaut class but wasn't selected. Still, David Leestma, who was the NASA director of flight crew operations at the time, took me aside and said, "Hey, Nicole, we didn't pick you this time, but we'd like to see how you do with crew operations, and see if maybe you make it next time around." With that, NASA offered me a job at the Johnson Space Center (JSC) in

Houston, in the aircraft operations group, as a flight engineer on the shuttle training aircraft (STA). This was the airplane they used to train astronaut pilots to land the Space Shuttle. Up to this point, for about ten years, I had been working as a NASA engineer in the space shuttle operations group at the Kennedy Space Center in Florida.

I was *very* excited about the STA flight engineer's job, which also included the opportunity to fly as the backseat crew member in T38 trainer jets. To qualify for that assignment, I had to pass the Navy's aviation water survival course, which included a swim test.

Three of us NASA trainees were inserted into a training schedule for about forty Navy cadets, though our activities were always kept separate. Clay Anderson had been selected for the NASA astronaut "Penguin" class of 1998 but had been unable to attend the training with his class, and Terry Lee, like me, was moving into a flight operations position. Shortly after our arrival at the Naval Air Station in Pensacola, Florida, the three of us waded into a large lap pool dressed in ill-fitting full flight gear that had been doled out to us after a quick visual "sizing."

The baggy green flight suit and beat-up helmet didn't concern me as much as the stretched-out, waterlogged combat boots, which were at least one size too big and impossible to lace tight enough to keep any water out.

The test required us to swim three lengths, in full gear. The first two lengths were in shallow water, about three feet deep. We were to start in one corner, swim straight ahead, then swim diagonal to the other corner, and finally swim straight down the other side to the deep corner (about eight feet deep), where we had to tread water while inflating our life vests.

All of this had to be performed without stopping, standing up, or grabbing the side of the pool. If I did any of that, I'd have to start over.

This test was the first thing we had to pass before we would be able to learn anything else, including helicopter rescue, parachute drag (how to handle being dragged by your parachute through water without drowning), zip-lining into the water while presetting all your gear to land safely, and lots of other pretty fun stuff. If I failed the swim test, though, none of that would matter because I'd be out of the course.

We stood in the water while an entire squadron of about forty cadets sat in the bleachers and awaited their turn. I have never been a strong swimmer, and I had already convinced myself that I was going to fail.

Clay went first, and it didn't help to watch him swim away. He made it look easy, like he was crawling across the top of the water and barely even getting his head wet. Then it was my turn. We could use any stroke we wanted but could take only one length of the pool on our backs—in case we couldn't make it any other way. I began with the breast stroke because Clay had made it look so effortless. Off I went. One stroke, two strokes, and then I began to sink. I wasn't freaking out at that point, but as I was moving my arms and legs, not much else was happening except the sinking. I'd move my arms and legs and sink, *glug-glug*. Over and over I sank and struggled just to get my head back above water and to catch a quick breath, so I could make some tiny progress forward. I'm not sure how I made it to the end of the pool that way. Thankfully there was no time limit.

For the next length I switched to the sidestroke, though I'm not really sure why. I only knew I couldn't continue with another length of flailing my way through the breaststroke, and I wasn't willing to switch yet to the last-resort backstroke. My sidestroke went about as well as the breaststroke, and I made it to the end of the lap only because I somehow managed to convince myself that I wasn't drowning. By this time, I was freaked out.

I'm sure the cadets sitting in the bleachers watching me were wondering why I was so freaked out by this silly swim test. I reached the end of the second length of the pool exhausted and questioning more than ever whether I would make it through the final length. I couldn't think of anything else to do but go for the backstroke, because standing up would mean I'd have to start over, and that, in my mind, was not an option.

Why on Earth had I thought the backstroke would be easier? That must have been my impression only because they said we could use it as a last resort. Well, it wasn't easier: as soon as I flipped over and started to move my arms, I began to sink. And now I was sinking on my back and struggling to pop my face up to the surface, so I could breathe. The damn boots were full of water and pulling me to the bottom.

Then, from a distance, I heard one of the young cadets yell, "Ma'am! Relax! Slow down! The helmet floats!"

I'm not sure how I heard him with all my splashing and mental screaming at myself, but he saved me with his simple words. I arched my back and laid my head back on the water's surface and basically used my floaty helmet all the way to the end of that final length of the pool. I have never been happier to get out of the water.

Not wanting my new friend to suffer as I had, I yelled to Terry, "The helmet floats!"

Later that day, in one of the classes before our helicopter rescue training, one of the Navy instructors used the "slow is smooth and smooth is fast" phrase. He pointed right at me and said, "Nicole, that's what you did on that final length in the pool." It's stuck with me ever since.

I am an optimist. Well, let's say I describe myself as an optimistic realist. I tend to believe that a solution exists for every problem.

During my NASA engineering days, one of my very favorite people on the planet, Jay Honeycutt, taught me one of the most important lessons I ever learned about problem-solving. Jay was the director of NASA's Kennedy Space Center. He is a hero of the Apollo and Space Shuttle programs and a trusted mentor and dear friend—Jay is like a second dad to me. At one of the first all-hands meetings with him after his arrival at KSC, he shared with all of us young engineers that we were in the business of finding solutions to challenging problems, and the way to do that was to always be thinking, *Here's how we can, not why we can't.*

These words, so simple and so powerful, have lived on a small sign on my desk and have stayed in the back of my mind ever since I first heard them. This idea also pairs beautifully with the "slower's faster" approach—use the best know-how we have to take action in response to the problems we face. We got to the Moon this way, and we can solve the problems here on our planet this way.

Even the staunchest optimist knows that there are things threatening our planet that we can't stop. For instance, in about one billion years our Sun will reach the stage in its life where it's so hot that it will cause all the water in Earth's oceans to boil and evaporate, leaving our planet uninhabitable.[2] A few billion years after that, the Sun will expand out to beyond the orbit of Mars and Earth will have disintegrated after being dragged into the surface of the Sun. Various scientific models portray different estimates of the relative timing and progression of these events, but this outcome has been broadly understood and widely accepted across the scientific community without controversy for more than one hundred years.[3]

We can't do anything to stop the evolution of the Sun, but we still might be able to save ourselves by becoming a multiplanetary species that can live beyond this solar system. Scientists and engineers are already investigating this possibility. We have some time to figure it out.

However, just as science says that our Sun will boil our oceans in about a billion years, scientists also have been warning us of a more immediate planetary challenge: the consequences of our own behavior. Science tells us that human activity is significantly damaging our planet's ability to continue to sustain life. The bad news is that we don't have one billion years to do something about it. The good news is that we *can*.

As human beings, we evolved to respond to immediate threats.[4] This makes sense: if we had not learned to run from predators, or move away from natural disasters, we wouldn't have been able to thrive as a species. We'd already be extinct.

But back when we were running away from predators, we had very limited access to information, and most of the information we got came firsthand, through our own direct experience. The problem is that in these modern times we're so bombarded with loads of (often contradictory) information from a wide variety of sources that it can be difficult to even see what's going on, never mind tell the difference between a real, immediate threat and a distant, "someday maybe" threat.

The same behaviors that help us respond to and survive immediate threats are working against us when it comes to a planetary challenge like climate change, which we have somehow convinced ourselves is not so "immediate." Our collective mind still hasn't figured out how to recognize that a planetary threat that is ten to thirty (or maybe even one hundred) years out *is* an immediate threat, even when the consequences of not responding fast enough could be cataclysmic.

I'm an engineer. I am not a scientist, but I played one in outer space. I've been aware for most of my adult life of the issue of climate change and other threats to our planet's life support system, and I've tried to consider these issues and threats with respect to my own actions and decisions. But I have often struggled to understand the science and to be able to identify which actions to take to make things better. My spaceflight experience made me even more acutely aware of our planet as our life support system, and I wanted to not only learn more but discover how I could be a better crewmate when I was back on Earth.

For this book, I did a lot of investigating and found information that helped me to better understand the threats to our climate, so I'll share my version of Climate Change 101 here with you.

Stepping back a little, to 1896, the Swedish chemist Svante Arrhenius predicted that burning fossil fuels would impact Earth's greenhouse effect (the natural process that warms the Earth's surface). Back then, it was believed that this warming from the additional carbon in the atmosphere might even be beneficial for future generations, but evidence quickly showed that the excess carbon dioxide released by burning fossil fuels would cause more harm than good. In 1975 the US-based geochemist Wallace Broecker coined the term "global warming" in his landmark paper "Climatic Change: Are We on the Brink of a Pronounced Global Warming?"[5] Thirteen years later, in 1988, I started my career at NASA as an engineer at the Kennedy Space Center, helping prepare Space Shuttles for their missions to space. That was the same year that Dr. James Hansen, a climate scientist at NASA's Goddard Space Flight Center, testified before the US Senate that "the greenhouse effect has been detected, and it is changing our climate now." He declared then, "with 99% confidence," that sharply rising temperatures were a result of human activity.[6] At NASA and also since retiring in 2013 and becoming the director of the Program on Climate Science, Awareness, and Solutions at Columbia University's Earth Institute, Dr. Hansen has devoted his work to understanding and protecting our home planet—as NASA itself stated in its original mission statement.

Hansen's 1988 testimony before the Senate is considered the first warning to a mass audience about global warming and the clear cause-and-effect relationship between excess carbon dioxide in the atmosphere and the frequency and intensity of weather patterns that are likely to increase and strengthen over time, with the potential to devastate our planet.

Later that same year, the United Nations formed the Intergovernmental Panel on Climate Change (IPCC) to provide governments and policymakers at all levels with regular scientific assessments of climate change—its implications, its potential risks—and to put forward adaptation and mitigation options.[7] Over twenty-five years later, in 2015, at the twenty-first session of the Conference of the Parties of the United Nations Framework Convention on Climate Change (UNFCCC)—more popularly known as COP21—the historic Paris Climate Agreement was adopted. Its aim was to halve greenhouse gas emissions by 2030 and reduce them to net zero by 2050—in time to limit the rise in global average temperature to no more than an increase of 1.5 degrees Celsius above preindustrial levels.[8]

Among the many voices speaking in support of the COP21 activities and the call to action outlined in the Paris agreement were astronauts. I was honored to be part of this group of international astronauts, which included two who were on the ISS at the time, to share a message of solidarity, hope, and cooperation to combat climate change with our political leaders. Our video message, titled "Call to Earth: A Message from the World's Astronauts to COP21," was played for all the participating leaders at the opening of their sessions.[9]

The Paris Climate Agreement is the most inclusive global accord on climate change to date—195 countries signed the agreement, and 189 had ratified it as of December 2020. While it doesn't bind any one country to any one solution, it focuses all countries on the same challenge of reducing emissions. "The Paris Agreement was built to be both flexible and resilient," said Maggie Comstock, a climate policy expert at Conservation International. "Countries and businesses will continue to take meaningful action because they get that climate action is not only smart for the planet—it can be smart for their businesses as well."[10]

I think for a lot of people, me included, it's hard to imagine how such a seemingly small change in average temperatures can trigger disastrous changes for life as we know it. Temperatures going up or down a degree or two in our everyday lives doesn't seem like such a big deal, so how could an increase of 1.5 degrees Celsius from where we are as a whole planet be so bad?

It's a little confusing because this 1.5 degree Celsius difference is not necessarily critical to humans on its own, but it is crucial to the operation of our planet, which we all depend upon to stay alive. All the natural systems critical to our survival—Earth's biosphere, geosphere, atmosphere, and hydrosphere—have one thing in common: heat is the factor that throws the delicate checks and balances of each system out of whack. This 1.5 degree Celsius change has been scientifically determined as the point where the climate impacts we're already seeing today will go from bad to significantly worse. I found one of the best ways to describe this on the website of the Climate Reality Project: "Another critical thing to understand about global warming is that it's not the case that everything up to 1.49999 degrees is rainbows and unicorns and free ice cream for everyone. (But once we cross the 1.5 degrees-line, the Four Horsemen of the Apocalypse polish off their martinis, look at each other, and say, 'It's go time.')"[11]

The more heat is added to the Earth's climate, the more out of balance the planet's life support systems become. The more out of balance they are, the more destruction and suffering we experience. Heat waves, more frequent and intense storms, melting polar ice, rising seas, coral reef destruction, whole ecosystems changing—these events will continue to intensify and interfere with our planet's life support systems, and *that* is why the threat we face is *immediate!* One of the other things I learned is that the increase of 1.5 degrees Celsius in the average global temperature so often referred to is measured from a baseline temperature set in the late nineteenth century, when the Industrial Revolution began and we humans started burning more fossil fuels. This is why you always hear about the change in temperature with respect to "preindustrial" levels.

Another important thing to note about global warming is that it doesn't affect every place on the planet in the same way. Some places on our planet, like the poles, warm much faster than others; for example, the Arctic warms faster than any other region on Earth, even if fossil fuels aren't burning there.[12]

So whether or not you believe in the terms "global warming" or "climate change" is not what's important. What matters is the scientific evidence that we humans have been negatively impacting Earth's life support systems for over a century and time is now running out. The damage to our life support systems will only get worse the longer we wait to change our ways. We're in a position now to reverse some of the harm we've caused by working hard to keep the temperature change below an increase of 1.5 degrees Celsius. This is what we need to do to ensure our future and avoid triggering some worst-case tipping points.

What is a "tipping point"? In the context of global warming, it's defined as a threshold where a tiny change could push Earth's systems dramatically and irreversibly into a completely new state.[13] Described very simply by Professor Tim Lenton, director of the Global Systems Institute at the University of Exeter, a tipping point is similar to the point where a child gently pushes herself over the top of a playground slide and finds it's too late to stop herself. Or it's like a game of Jenga—you remove block after block from the tower and place each one on top until the tower can no longer support itself and tips over.[14] The time frame for the Earth's response to human activity and climate change and the time frame for the child's ride down the slide or the Jenga tower falling might be different, but the feature they have in common is that once the collapse has started, it is virtually impossible to stop.

While the precise levels of climate change sufficient to trigger a tipping point remain uncertain, we know that the risks associated with crossing multiple tipping points increase with rising temperatures. Scientists have identified nine tipping points that could be triggered by the increasing temperatures associated with climate change. One in particular that caught my attention is melting permafrost. "Permafrost" is the name given to ground—soil or rock—that contains ice or frozen organic material that has remained at or below freezing temperatures for at least two years; in fact, much of the planet's permafrost has remained frozen for hundreds, thousands, and even tens of thousands of years. Permafrost covers around one-quarter of the land in the Northern Hemisphere—about nine million square miles, including large expanses of Siberia, Alaska, northern

Canada, the Tibetan plateau, and other higher mountain ranges—and it varies from a few feet to a few thousand feet in depth. Permafrost is also found in the Southern Hemisphere in parts of Patagonia, Antarctica, and New Zealand's Southern Alps, and subsea permafrost occurs in shallow parts of the Arctic and Southern Oceans.

"So what?" you might ask. "I live in Florida. How could changes in permafrost have anything to do with me?" Well, it does, and it's scary. Like everything on the planet, it's all interconnected. When permafrost is frozen, it's harder than concrete, and it functions as a "sink," meaning it absorbs and stores significant amounts of greenhouse gases. When it thaws, scary stuff happens.

Permafrost holds a vast amount of carbon—around 1,500 billion tons are stored in the Northern Hemisphere. That carbon accumulated from the burning of fossil fuels and from dead plants and animals decomposing over thousands of years. That means around twice as much carbon is stored in permafrost as currently exists in all the Earth's atmosphere. When the permafrost melts, everything falls apart. It's like a prehistoric-sci-fi-horror film as buildings, roads, and towns sink and collapse. In Russia, ruptured fuel tanks have dumped tens of thousands of tons of oil into the surrounding land and waterways; sprawling boreal "drunken forests" are sinking into the softening ground below; lakes are draining; seashores are collapsing; and whole ecosystems are radically changing. So far, the most impacted have been indigenous communities and the wildlife that depended on those ecosystems and have now been displaced and forced to adapt not only to a new home but to a whole new way of life.

Now top it all off with two really scary twists. As the permafrost melts, instead of serving as a massive natural sink for carbon, it becomes a massive *source* of carbon as it releases all the carbon dioxide and methane it's been storing. And if massive amounts of greenhouse gases escaping into the atmosphere isn't scary enough, brace yourself for the release of the ancient bacteria and viruses that have been trapped in that same frozen ground for thousands of years. Scientists have discovered microbes over 400,000 years old in thawed permafrost. These newly thawed microbes have already made some modern-day humans and animals very sick. According to one news report,

> In the summer of 2016, a heat wave washed over Europe, thawing permafrost in the north. In the Arctic soil of Siberia, bacteria began stirring—anthrax, to be specific. The thawing, shifting ground exposed a reindeer carcass buried and frozen in 1941. The anthrax spores from the body found their way into the top layer of soil and the water nearby, before being picked up by thousands of migratory reindeer grazing in the area. Over two thousand reindeer soon contracted the deadly bacteria and passed it along to the nomadic Nenets peoples who travel alongside the reindeer and depend upon them for food. By the end of August, a 12-year-old boy had died, and at least 115 others had been hospitalized.[15]

As the Arctic warms twice as fast as the rest of the world, and with the permafrost already beginning to melt, scientists predict that approximately 40 percent of the world's permafrost could disappear by the end of the century. Warmer conditions cause the release of more carbon dioxide and methane from permafrost, which means more warming, which in turn causes more permafrost to thaw, and so on—a vicious cycle. The tipping point would occur when the permafrost has melted to the point where more greenhouse gases are being released into the atmosphere than we could ever pull out of the atmosphere to counteract it.

This is absolutely scary stuff. But what I love is that the scientists still have hope, which means that we should have hope—hope enough to take action. The hope is in the fact that solutions exist. It's up to us to take the steps necessary to radically reduce our human impact on the environment, and as humans, we are uniquely capable of finding even more solutions and developing more technology to prevent further impacts. We just have to *do it.*

NASA satellite and airborne systems are helping scientists better understand the changes in our atmosphere and on the surface of Earth. The Soil Moisture Active Passive (SMAP) satellite system maps soil moisture everywhere on the Earth's surface and detects whether soils are frozen or thawed; these observations can help scientists understand where and how quickly the permafrost is thawing. The Orbiting Carbon Observatory (OCO) is a carbon dioxide measurement system located on the ISS; it can create mini-maps of carbon dioxide concentration over places of interest on the planet and help us better understand the sources and sinks of the greenhouse gas.[16]

The Arctic Boreal Vulnerability Experiment (ABoVE) is an airborne system that measures atmospheric carbon dioxide, methane, and carbon monoxide and collects land surface data to improve our understanding of the carbon cycle in the Alaskan Arctic. The measurements are used to enable accurate forecasts of how much carbon dioxide and methane will be emitted into the atmosphere as permafrost thaws.

If we want to survive—better yet if we want to thrive—we must continue to increase our understanding of the impacts we're having on our planet and do everything we can to keep global warming below 1.5 degrees Celsius. The good news from the scientists is that we still have time to do so, but we need to make haste! Compare our current climate change condition with an emergency on the ISS. The first thing we need to do as the crew of Spaceship Earth is diligently and deliberately run the boldface steps needed to stabilize our environmental situation—then we'll have the time to "go slow to go fast" and deal with the longer-term underlying problems. Fortunately, we humans have an extraordinary capacity to solve complex problems, even large-scale ones that we've caused ourselves, like climate change.

So what do we need to do? The Intergovernmental Panel on Climate Change's recommendation is that we "cut global greenhouse gas emissions by 50% by 2030 in order to limit temperature increase to 1.5°C." I understand what the IPCC is saying, but until recently I had no clue what that meant for me individually, or what I could do to help make the necessary changes.

As I detailed earlier, on the space station we have a boldface checklist that specifies the immediate steps to take in the event of an emergency. Where is the boldface checklist to address climate change? Imagine if the alarms were sounding on our space station and we decided to use our limited time to respond by asking one another if we think it's real or not? Or by wondering if we even should respond? Or by getting irritated by the sound of the alarm and upset because no one else is silencing it?

This makes me think again of the Navy swim test. We were thrown in the pool with no preparation or training for how to successfully complete the test. We were left to flail and figure it out on our own. For the most part, that's what we're dealing with here on the planet. We've all been born on this planet that supports our lives in the deadly vacuum of space. We've operated in a way that has brought much comfort and improvement to our way of life, but also great threats to our survival.

With no operating manual, no boldface checklist handed to us, and no training, it's like humanity is flailing in the Navy training pool. In the case of climate change, science is the Navy cadet yelling to us what we need to do.

My flailing in the pool stopped when I heeded the words of the cadet. It wasn't comfortable to arch my back and push my head back in the water and use the helmet to float. Nothing about it seemed natural or like something I would have known on my own to do, but it worked! It helped me complete the test and escape the threat of drowning in the water.

Reducing emissions is like a floating helmet, and we've got to lean back and kick. It won't be comfortable, it might not even come naturally to us, but it is the thing we need to do to survive. Sometimes, to borrow another Navy motto, "the pleasure is the pain."

The climate science presented in the IPCC report provides us with our mission—"reduce global greenhouse gas emissions by 50% by 2030." But how?

I've done a lot of research in my pursuit of ways to change my own behaviors and guide my own actions in order to reduce my carbon footprint. It was reassuring to discover many overlaps across the different sources I consulted. One book in particular stood out for me, *Drawdown: The Most Comprehensive Plan Ever Proposed to Reverse Global Warming*, by Paul Hawken and the Drawdown Project.

Drawdown helped me identify changes I could make in my life that I feel certain are meaningful in the tactical scheme of a personal boldface checklist. At the higher level, *Drawdown* calls for four connected actions: reduce emission sources, protect greenhouse gas "sinks," improve society by fostering equity for all, and be an informed and active citizen. The actions that have the most potential in each of these four areas are what I would call our boldface checklist. In the following, I've added the actions taken by my family and I so far in response to this call for action:

Reduce emission sources. The actions identified as having the most potential for reducing emission sources are shifting production of electricity and energy away from fossil fuels, improving energy efficiency with LED lighting, reducing food waste, shifting to plant-based diets, monitoring and repairing refrigerant leaks, shifting away from daily commutes and toward working from home, using public transportation, driving electric vehicles, and improving insulation in buildings.

Although we don't have the ability to shift how energy is produced for our own home yet, our family has chosen the service plan from our electric company that relies the most on renewable sources. We have updated all lighting in our home to LED. We have modified our grocery shopping habits to commit to eating all that we buy by purchasing only the food we need when we need it, and we have changed to a vegetarian diet (trying for vegan). We have biannual checks done on our air conditioning and heating system to make sure it is operating properly and there are no leaks. We are in the process of updating the insulation in our home. We are fortunate to have been able to purchase an electric vehicle and have reduced our transportation to just one car.

Protect greenhouse gas sinks. Greenhouse gas sinks are the natural systems that suck up and store carbon dioxide from the atmosphere. Land is the major sink. The actions identified as having the most potential for protecting greenhouse gas sinks are agricultural practices like silvo pastures that integrate trees, pasture, and forage into a single system; restoring tropical forests; and planting tree plantations on degraded land.

With respect to sinks, we have designed the yard around our home with native plantings, a small herb and vegetable garden, and as many trees as possible. We support local community gardens and restoration activities as well as global organizations that are working to restore tropical forests and coral reef ecosystems.

Improve society by fostering equity for all. The premise here is that, by moving beyond fossil fuels toward clean energy and the climate solutions associated with emission reduction and protection of sinks, we will directly improve the lives of people around the planet, including those in the most remote and impacted regions. Increasing accessibility to more and better education for girls and young women has been identified as one of the most effective ways to improve society and foster equity.

We are working with schools and organizations, both locally and around the world, to encourage the education of girls and young women. As a retired astronaut, I am fortunate to be able to share my experience with girls and young women in a way that I hope is inspirational for them and that encourages them to pursue exciting and challenging careers, and to believe they have the power to make a positive impact in the world.

Be an informed and active citizen. We also want to make more informed choices when we vote because we know that, beyond our personal actions, our right to vote gives us the most power to effect change. This means not just checking one box and voting for everyone in your political party, but considering each candidate on their individual merits.

Selecting a good crew means seeking out the best candidates for the mission. Imagine, for a minute, that when you vote you're picking your crewmates on the space station. When I do this, I ask myself, *Is this a person I would trust to be accountable for doing the kind of job they said they would do? Someone who would have my back in a crisis situation?*

I recommend you think of voting in this way because whether you realize it or not, this is in fact the choice that we are making. Only we aren't on a space station, we are on Spaceship Earth, and here there's less built-in accountability, so we have to create it. We can only do that by voting and then paying attention, staying engaged, and demanding that our elected officials and policymakers—our representatives—follow through with the changes we elected them to make.

It's been encouraging to see the positive impact of the personal lifestyle changes we've made at home not only on our life as a family but also on the lives of the people around us. We are trying to do our part by making these changes at home, and it's a great start, but I now realize that it's not enough. We need to continue to push ourselves and go beyond our own backyard. My family and I want to become even more conscientious consumers, so we're focusing on consuming with conscience and consideration for where the things we buy come from, how they're produced, and where they'll go at the end of their useful life. Beyond meeting our own needs, we want to choose what we consume in a way that supports a sustainable future in line with the necessary changes we need to make in response to climate change.

The initial effort it took to make some of the changes in my own life has now become easier, and it makes me happy when I think about the good that's coming from making these shifts. I'm in a place now where I better understand how my own actions can positively influence life around me. And I am glad I get to share what we've done in the hope that others will be encouraged to make some changes and experience the goodness too.

In a 2018 article about the progress associated with the work of the IPCC, one of the scientists who participated in the work, Paul Valdes from the University of Bristol, stated, "It has shown that human induced climate change is not an opinion but a scientific fact." He added, "I would strongly argue that the barriers for progress are no longer scientific."[17]

It's now a matter of summoning the will to make progress happen in line with the goals of the Paris Climate Agreement—to cut fossil fuel emissions in half by 2030. Local, state, national, and global policy changes are necessary across multiple areas, including shifts in personal choices, individual actions through voting, and greater political engagement. We all need to find our own way to become part of the solution and to encourage action, starting in our neighborhoods and then spreading throughout our planetary community.

There's an enormous difference in how our planet will look if we choose to cut back on greenhouse gases and how it will look if we don't. I really like the way the IPCC reports use simple graphs to show us this difference—a red line representing no change, and a thin blue line that shows our potential if we do. In September 2019, at a presentation of the IPCC findings at the Explorers Club in New York City, the scientists stated that "the blue line is where there is hope."[18]

I'm hopeful because the scientists show us there is reason for hope. I'm hopeful because we have a solution. I'm hopeful because we *are* capable of choosing to implement the solution, no matter how challenging it might be. I'm hopeful because science tells us that we still have time to make the difficult choices that will move us in the right direction. And I'm hopeful because I see many different organizations and people (including those showcased in this book) who are responding to these epic challenges. They aren't waiting for someone else to do it for them. They're taking action and doing what they can to encourage action from individuals, organizations, governments, and the business community.

Because the "I" in "ISS" stands for "International," the majority of my training in preparation for my ISS flight was held outside the United States at our international partners' sites around the world. Over 50 percent of my time during that three-year training period was spent in Russia, Japan, Europe, and Canada, and I necessarily also spent a lot of time in the seat of an airplane flying to these places and back. My routine for these long flights included stowing my computer and a pen in the seat pocket for later use and plugging my phone in to charge. Everything else went overhead. After getting situated in my seat, the first order of business was to pick up the airline magazine and go straight to the back for the crossword puzzle and sudoku, with the goal of completing those before takeoff. Then sleep.

In 2017, on a Delta flight, I opened the issue of the airline's *Sky* magazine entitled "Project Earth." The cover picture was of a tall man about my age standing outdoors, in a field, with his hands on his hips, with windblown hair, and wearing a confident smile and a black long-sleeved shirt under a zipped-up silver Patagonia vest as he looked off into the distance. Behind him was a waterway and the almost imperceptible silhouette of some kind of industrial plant on the horizon. The cover's subhead read, "The Nature Conservancy CEO Mark Tercek's mission is to convince the world that environmentalism is good for the planet—and the economy."

This gripped my attention so firmly that I skipped the crossword and sudoku and went straight to the article, which was entitled "The Call of the Wild." According to the table of contents, "Once an investment banker, Mark Tercek is now the celebrated president and CEO of The Nature Conservancy (TNC)—and he's redefining the intersection of business and environmentalism." I was so enthralled that I read the article several times. I didn't even notice when we took off.

Underscoring the idea that we should be "protecting nature *for* us, not *from* us," Tercek's perspective on bringing business and environmentalism together seemed so perfect, and even simple, that it surprised me that businesses around the world hadn't already adopted his approach. I remember chuckling that it seemed a lot like the so-called divide between science and art, or the left brain and the right brain. (News flash: we need to use our whole brain!) After reading the article, I thought more about the need for all of us to focus on finding common ground if we are to overcome our greatest challenges. I found myself looking for ways in my own life to recognize this common ground, professionally and personally, and through the work I was doing to bring art and science together. I also noticed that I started sharing what I'd learned about Mark Tercek's philosophy. (For more insight into Mark's philosophy, I highly recommend his book *Nature's Fortune: How Business and Society Thrive by Investing in Nature.*) I became hopeful that I'd have the chance to meet him in person one day.

That day came about two years later at TED2019 in Vancouver. Having participated myself as a speaker at a couple of different TEDx events, I decided it was time to finally experience the live presentations and overall atmosphere at the big TED conference. My husband Chris and our dear friends Michael and Margaret Potter had been attending for years, but I'd been a bit skeptical of the return on investment. On day one of my first TED conference, the return on my investment was realized.

We had just signed in and were standing in the "goody bag" line. I turned around and stood face to chest (he is quite tall) with Mark Tercek. I actually took a step back to confirm and then quickly turned back to Chris (who knew of my interest in Mark's story) to whisper to him who was behind me. He told me to introduce myself. Since TED is all about networking and meeting people to help make a positive impact in the world, I got up the nerve and introduced myself, despite feeling like a total fangirl. It went something like this: "Oh my gosh, you're Mark Tercek! I read an article about you in an airplane magazine a couple years ago and have been following you ever since. Following, not in a stalky way (ha ha... awkward).... Thank you for all you're doing... so nice to meet you.... I'm writing a book... maybe I could interview you sometime...."

Despite my wacky introduction, he was very kind, and we established contact on the TED app. Honestly, I thought he'd never want to speak with me again.

Fast-forward to a dinner event that night hosted by one of my husband's friends, and can you guess who the host sat me next to? Yep! I was much calmer this time and by the end of the evening felt that I might have redeemed myself through our dinner conversation. Mark told me he looked forward to hearing from me about my book. On the final day of the conference, we watched him in action as he presented the Nature Conservancy's "Audacious Plan to Save the World's Oceans."[19]

As promised, Mark did allow me to interview him for this book (a few times actually). He is a self-described nerd and a late-blooming environmentalist. "I grew up in Cleveland," he told me. "I was outside all the time, but it was in a grimy city, so I was not an environmentalist or a nature boy."

"But when I was a twelve-year-old, in June 1969, my river caught fire."

When he said, "My river caught fire," I flinched. Had I heard him correctly? A river (his river?) caught fire? I never heard of any river catching fire!

Turns out that when the Cuyahoga River (Mark's river) caught fire in 1969, it was not the first but the thirteenth time it had happened. The first time this river caught fire was back in 1868, but owing to the cleanup efforts sparked (pardon the pun) by the 1969 fire, that would be the last.

Back in 1969, the Cuyahoga River smelled of sewage and was a dumping ground for industrial waste. Nothing had lived in the river for decades, though history shows that before rapid industrialization in the late 1800s the Cleveland waterways were "some of the most pristine ever seen."[20]

The 1969 fire started when a train crossing on a bridge dropped a spark onto the surface of the water, which erupted into flames five stories high and spread across the width of the river. At that time, two-thirds of US lakes, rivers, and coastal waters were unsafe for fishing or swimming. This fire became a "potent symbol for the nascent environmental movement."[21]

Cleveland's mayor, Carl Stokes (the first African-American mayor of a major US city), held a press conference the day after the fire and requested help from the state government to clean up the river. His presence at the site of the fire helped transform what might have remained a local event into national news. Several other major environmental catastrophes had occurred in the United States earlier that year, including a massive oil spill off the coast of Santa Barbara, so this one prompted more outcry for cleanup and reform from environmentalists, and *Time* magazine ran an article about the Cuyahoga River fire.

One month later, Neil Armstrong and Buzz Aldrin walked on the Moon, the first Earth Day was held ten months later, and within three years the government had taken meaningful action by forming the US Environmental Protection Agency (EPA) and passing the Clean Water Act.

The Cuyahoga River has come back to life—some even refer to it as the Lazarus River. Interestingly, the river still has some dangerously high levels of bacteria, but even with this issue to be managed, it is significantly improved and unrecognizable as the river that caught fire in 1969. The city of Cleveland continues the remediation work on the river and is also one of four hundred cities around the country that, despite US withdrawal from the Paris Climate Agreement under the Trump administration, has committed to the agreement's goals. (At the time of this writing, the United States had rejoined the Paris Agreement.) These cities are aiming for an 80 percent reduction in greenhouse gas emissions by 2050, as well as 100 percent of their electric energy coming from renewable sources.

The first sentence of the *Sky* magazine article about Mark Tercek reads: "There is an inner environmentalist in us all, poised to burst into bloom." Mark returned to this idea several times during our interview, and not just in reference to himself. "There's an inner environmentalist in everyone," he said. "I'm like exhibit A. It was quite late in life when I became interested in nature. Then the businessperson in me said, 'Wow, we don't want to spoil nature or ruin it. How can all my problem-solving skills address environmental challenges?' That's how I became an environmentalist, and lots of people can follow that pathway. It's kind of a geeky pathway, though. It's not like the Earthrise moment, but it works."

Even as a child in Cleveland, Mark knew that a river on fire was not okay. He was very clear in the interviews that neither he nor his wife grew up immersed in nature, but when they became parents they felt that it was important for their children to have the opportunity to establish their own personal relationship with nature. As a family, they started to spend their weekends and vacations at nature-oriented locations. Mark described to me one of their trips, to Costa Rica, in what I would call a very "Earthrise moment" way.

"We were roughing it on this trip, literally really in the mud. We had an extraordinary guide who helped us appreciate the wonders of Costa Rica's ecosystems... something really clicked for me there, and I think for my family too, like we woke up together to the wonders and joys of nature."

At the time when Mark was discovering "how fantastic it is to spend time in the great outdoors," he had already been working for over twenty years as an investment banker for Goldman Sachs. "The Costa Rica trip ignited a love for nature that inspired the geek in me to learn more about the way our ecosystems function and the kind of impacts we have on them."

He made sure they had guides who had the expertise to not only provide a great experience for their family trips but to also share the awe and the challenges of the natural environment. He started donating to various conservation groups, reading environmental literature, and contacting environmental experts for further discussion about what he'd read. One book in particular resonated with him, *The New Economy of Nature: The Quest to Make Conservation Profitable*, by the Stanford University ecologist Gretchen Daily and the journalist Katherine Ellison. Their book presents ecosystems in terms of the goods and services they provide to humanity and the profit that can result from protecting them. Mark's immersion in nature itself, combined with what he was learning about the value associated with conservation, inspired him to speculate on how he might use his Goldman Sachs business experience to "make the case for why protecting nature is the smartest investment we can make."

He returned to Goldman Sachs with this new motivation in mind and made his case. Mark went on to develop the firm's environmental strategy, and in 2005 he launched its Environmental Markets Group (EMG). The EMG establishes partnerships with corporations, academic institutions, and nongovernmental organizations (NGOs) to develop innovative market solutions that address environmental and social challenges and to arrange large-scale financing and investing that will help scale up clean energy markets and support long-term sustainable economic growth and the transition to a low-carbon economy.[22]

Hoping to do still more, in 2008 Mark became the CEO of The Nature Conservancy (TNC), which he helped to adopt a more businesslike approach as he championed its programs to promote nature as a capital good worthy of investment. After eleven years leading TNC and helping to reshape the approach of other conservation groups, Mark moved on in July 2019 and now advises established companies, start-ups, institutional investors, and NGOs on environmental and organizational strategies and on opportunities in impact investing, which supports social and environmental causes while also generating financial returns. He is recognized as a global conservation leader and a champion of the idea of nature as capital, a concept he describes as "valuing nature for its own sake as well as for the services it provides for people, such as clean air and water, productive soils, and a stable climate."

Mark discovered his inner environmentalist and is continuing his mission to help us all find and harness the power of our own inner environmentalist. "When I joined TNC," he told me, "one of my jobs was to go to all the events where we're thanking our supporters for being so great, and my wife Amy would usually join me. A lot of the donors and supporters are elderly people. Not all of them, but a lot of them are in their late eighties. So for the first time in my life I began to spend my time with older people, and I was blown away. They were highly engaged. They were dynamic. They seemed almost youthful. Amy and I commented to each other how it was puzzling—they're the old people, but seem like the young people."

As Mark and Amy got to know these people better—many would become good friends—they figured out the puzzle. "They decided to find a cause they cared about," he said. "In this case it was the environment, but it could be any cause. They give it their all. So right off the bat, one big thing happens. They're not sitting worried about themselves. They're thinking about others. And they're busy. They're working with young people. They're working on how to solve problems. These people in their late eighties, they were even taking care to take good care of themselves, to stay in shape so they could do this. There you have it," he concluded, "the secret to a happy life!"

As Mark shared this discovery of the joy of living a life of purpose, I described for him how living as a member of a space station crew is also driven by purpose and service, and also brings joy.

"I'm going to steal that term from you if you'll let me," he said. "I can say I have this friend who's an astronaut, and I was telling her about how I think being engaged in causes can lead to a happy, fulfilling life, and you said, 'Yeah, it's like being a crew member.' I think that's important because there's a lot of anxiety and unhappiness out there."

As Mark and I continued to talk about this idea of behaving like crew, the conversation turned to the reality that, as a whole, we are not living like crew here on Earth. We are not working together to operate our spaceship's life support systems for the benefit of all life and the planet. Mark told me that what I related about living like crew made him think about how much we Earthlings are not like the inhabitants of a spaceship. "You guys [astronauts] got organized. You got selected. You got trained. I presume if you're an astronaut you have to know how everything works even if you're not the expert.... You're in your spaceship, and you're in trouble, and you just decide to blame someone.... Guess what? That doesn't fix anything.... You're kinda stuck with one another. You gotta work together and obviously, on our planet, we're stuck with one another.... I think your bit about being a crew member versus being a passenger is right. Being a passenger, just going for the ride, is a pretty lousy way to live a life.... It's really quite different here from your spaceship, I think."

The analogy of Earthlings embracing the value of our role as crewmates here on Spaceship Earth is one that resonates with everyone I speak with, and it is the central tenet of the *Back to Earth* story. While the human species still hasn't figured out how to live like a crew of Earthlings, many are making the effort to do so. One example is the group of older people Mark encountered who are finding their joy in life as TNC crew members. His own work in showing us how business and environmental motivations can come together for mutual gain is another, as is his discovery of his inner environmentalist. These are all ways that the value of "living like crew" can be seen in the life of just one person.

I believe that all of us are poised (perhaps without even knowing it) to burst into bloom as Earthlings. We just need to decide to do it.

Mark and I spoke about how important it is for all of us to find meaningful changes we can make in our daily routines that can help offset the rise in greenhouse gases and extend the viability of our shared life support system here on Earth. But if we really intend to turn around climate change trends, individual changes alone won't be enough.

"The most important thing for environmental progress is government policy," Mark said. "So, for each of us to help influence government policy, voting is by far the most important thing you could do. Everyone has to get more politically motivated and engaged."

"Sadly," he continued, "the numbers show that relative participation in voting is low, even among the environmental supporters and the younger people who are voicing the most criticism and calls for action. Engagement doesn't mean criticism. Don't just criticize the politicians. Learn more about who they are and their policies and vote for the ones that will make the change. Then hold them accountable."

In November 2009, my American crewmate Jeff Williams and I voted from outer space. We set up little voting booths and sent pictures of ourselves casting our ballots to our county polling office in Houston. The elections office staff greeted the pictures with surprise, excitement, and gratitude that we would take the time to share our experience with them.

I am thankful to have been able to exercise my *right* to vote during my first spaceflight. It made me realize what a privilege having the right to vote is, and how important it is to make voting a priority and to make it accessible to all our citizens, no matter where they are.

NASA astronauts didn't always have the ability to vote from beyond the atmosphere. Nothing was done about it, however, until John Blaha raised awareness of the fact that he missed out on voting in the 1996 presidential contest between Bill Clinton and Bob Dole because he was living at the time on the Russian Mir Space Station. When Texas state senator Mike Jackson heard about it, he went to NASA and told them he wanted to introduce a bill. Jackson had won his election in 1989 by just seven votes out of twenty-six thousand, "so I was all in favor of letting more people vote," he said. Other legislators from both parties were quick to join him. Now retired from the Texas legislature, Jackson reports that his colleagues' first response was "You want to do what?" But they soon realized, "Why didn't we think of this before?"[23]

Thanks to the legislation in Texas, where most astronauts live, the first astronaut to vote from space was Dave Wolf in 1997, during his stay on the Mir space station. Dave doesn't recall what was on the ballot that year, but he remembers how moved he felt.

"I voted alone up in space, very alone, the only English speaker up there, and it was nice to have a ballot in English, something from America," Dave said. "It made me feel closer to the Earth and like the people of Earth actually cared about me up there."[24]

The process for voting from orbit remains pretty much the same today—and it's pretty easy. Mission control sends astronauts an encrypted email containing an absentee ballot. Astronauts fill it out and email it back to Earth, where it reaches a county clerk, who opens the file and writes in the selections on a paper ballot. Only that clerk knows how an astronaut voted.

When astronauts get their absentee ballots via their email in space, their address is listed as "low-Earth orbit."

Sometimes the only way to get action in Washington is to have enough citizens voting for change and forcing Congress to put people's interests ahead of business. The words "everyone needs to vote" were said multiple times by Mark Tercek during our interview. In fact, he was a lot more direct about it a couple of times, saying, "Vote, damn it!" He was impressed that astronauts had not only considered the importance of their vote even while they worked as the crew of a spaceship on a mission of science and exploration, but that they pushed for the right to vote from space. When I asked him how we will build our Spaceship Earth crew, he replied, "The two most important things are to vote and to find common ground with each other."

The burning river in Mark's hometown of Cleveland has been cleaned up owing to government policy (the Clean Water Act), and the air we breathe is less polluted because of action taken in response to government policy (the Clean Air Act). Government policy drives changes in the ways we do business. Sometimes the only way to motivate industry and government to take the necessary action is for enough citizens to vote for change and force Congress to put the long-term interests of humanity before interests focused on short-term gains.

"I really like the clean air story," Mark said (several times). "When the Clean Air Act was proposed, it was ferociously opposed by the auto industry, who said, 'Oh, that's impossible. It'll cost billions of dollars, billions in jobs.' But the air quality was so bad, amazingly, Congress passed the bill. So then what does the auto industry do, since they have to comply with these laws? They invent the catalytic converter. Voila! They achieve the tailpipe emission reduction, factor the expenses, and that's why we have clean air today."

"Business is a source of innovation," he continued, "but no way they were gonna do it on a voluntary basis. They had no choice but to comply. They invented the technology that let everything come together. I like the clean air story."

"On the climate front, it's chaos," Mark said. "We just don't have very many rules for it. Like right now the rules allow for us to do stupid things like dump carbon at zero cost. How about we start with a policy that makes it illegal or very expensive to emit carbon pollution?"

"This sounds like a reasonable policy to implement," I said. "It sounds like a 'Clean Air Act' approach to carbon, where a change in policy could create the force needed to drive the necessary changes. While policy change can be difficult, this is something worth working on together."

"One of the things I've learned," Mark told me, "is that we sometimes make a big mistake: we presume that the people who don't see the world the way we do are bad people or they're opposed to what we want, or they're stupid, or they're greedy. There are some people like that, so you can't dismiss that, but I think we overdo it. Mostly we need to be willing to show some interest in their perspective. Even if you don't agree, you need to put that aside for a moment and try to have earnest, respectful dialogue, find some common ground. Find some things you can work on together and take it from there. I think it's an enormous untapped opportunity."

Many, including Mark, argue that climate change is "the hardest thing humankind has ever had to address." Thankfully, humankind has a history of finding solutions for extremely challenging and complex problems. Underlying all these solutions, in one way or another, was a decision driven by some greater purpose. Using the example of the International Space Station, one of humanity's greatest engineering challenges, that greater purpose was to work together in space to improve life on Earth. That greater purpose, the common ground we've found in partnership with fifteen other countries, is what brought us together in the first place, and it's what we base our decisions on every day and what leads to our mutual success. By looking for the common ground, we are agreeing that we believe it's important to find a solution.

Imagine if we just looked at climate change (or any other challenge) from the Earthrise perspective that regardless of our external differences, we all are on the same spaceship. As Buckminster Fuller said, "Now that we have a way to visualize the whole planet with greater accuracy, we humans will be better equipped to address challenges as we face our common future aboard Spaceship Earth. We must learn to see what unites us rather than what separates us."

What unites us is that we all live under the same sky, we all see the same Moon, and we all have the same basic need to survive. What unites us is that we have the ability and opportunity to respond to our greatest challenges in precise and deliberate ways. What unites us is that we *can* work together to ensure the sustainability of life on Earth.

We find ourselves today in a situation where response time is critical. We've been so good at "going slow" and assessing the problem that we've let the time for going slow elapse, and now it's time to go fast. We know the boldface checklist, and it must be run quickly to restore the integrity of our life support system. On a space station, the purpose of a boldface checklist is just to get the situation stabilized so that the station and everyone on it is safe. Then the hard work begins of figuring out the cause of the emergency and the safeguards to put in place so it doesn't happen again.

Here on Earth, alarm bells have been going off for decades, and we are still tinkering with just the idea of executing a boldface checklist. We've gone slow for too long, and it's time to get moving. We need to execute the boldface plan and take the actions to mitigate climate change that we already know are required. Only then can we "go slow to go fast" toward a shared future where we as humans can learn to behave as Earthlings in partnership with all life on our planet, so we can all thrive together.

I've worked with children around the world through my Space for Art Foundation. At one pediatric cancer hospital in Texas, I met a young girl with her sights set on the stars.

NASA

CHAPTER 6

STAY GROUNDED

"YOU ARE GROUNDED." These are both the worst and the best three words you can say to an astronaut. On the downside, to be grounded from spaceflight means that something has happened that put you on the "Do Not Fly" list. On the positive side, being a grounded person is a vital attribute to ensure an astronaut's success.

One of the greatest compliments I receive is being described as "down to Earth," though the person telling me that usually chuckles because they think it's funny to describe an astronaut that way. In fact, people seem surprised that an astronaut would be so grounded. I can't help but think about the absurdity of that perception. I mean, can you imagine what a disaster it would be if you had a bunch of "space cadets" operating a space station? That could only work in a spoof film, perhaps a next-generation version of *Airplane!*

To be grounded is integral to all the ways in which a successful crew operates, as described in the previous five chapters. As a crew, we operate with the perspective that everything is local. We are acutely aware of the integrity of our thin metal hull—our spaceship's equivalent of our planet's thin blue line. We embrace the value in the diversity of our crewmates and what that diversity brings to mission success. We understand that none of us can afford to behave like passengers. And from the beginning, we approach our missions with the proper planning that allows us to go slow to go fast. Mission success depends above all on a well-balanced and sensible crew who all understand that their actions must be based firmly on evidence, who are present to and respectful of one another, who are situationally aware, and who have trained their minds to stay relaxed and calm so that they can respond to whatever is going on around them.

Just as our crew must be stable, safe, and grounded, so must the ISS itself. As our temporary home in space, the ISS is essentially a football-field-sized conglomeration of high-voltage, ginormous, positively charged solar panels and a strategically configured metal structure with big boxes of electrical equipment, moving in unison through the negatively charged particles of the surrounding ionospheric plasma—a unique characteristic of the station's place in space. Orbiting above the planet at a roughly 250-mile altitude, the ISS is smack dab in the middle of the ionosphere, which overlaps the top of the atmosphere and the beginning of space. It's named, appropriately, for all the negatively charged particles (ions) circling Earth at this level. Thus, like our homes on Earth, the ISS needs to be electrically grounded to make it safe for its crew to live there in space.

Electrical grounding balances out the negative charge with the positive by acting as a "safety valve" that releases excess charged energy, giving it somewhere safe to go other than through us astronauts. This significantly reduces our risk of dangerous electrocution or fire. It also protects the electrical equipment from being damaged by excess current. For our homes on Earth, the excess charge is typically dispersed through a wire running from the electrical equipment to a conductive metal rod in the ground. Earth, as it turns out, is a good conductor, and large enough to accept or supply an unlimited number of charges without its own electric charge getting out of balance.

Fortunately, there are many different ways to ground an electrical system, and some don't even require that direct physical connection to Earth. That's a good thing for us on our spaceships because it wouldn't be possible to connect a 250-mile-long wire between our space station and a metal rod stuck in the planet below.

To electrically ground the ISS, we use a piece of equipment that sounds like something right out of *Star Trek*—the Plasma Contactor Unit (PCU).[1] The PCU creates a stream of ions and electrons that neutralize the positive charge and carry it safely away from the structure and into space. This "grounding" of the ISS allows the electrical systems to operate effectively and efficiently, and it enables the crew to operate safely, both day-to-day inside the space station and outside during a spacewalk.

A grounded crew and a grounded spaceship provided the safety, security, protection, and comfort I needed to feel that the ISS was my home in space. My crew was like family. This experience expanded my understanding of home and redefined what it means to me.

When I first got to space, I would float to the window with excitement every time I knew we were over Florida. I considered Florida my home. I wanted to experience Florida from space.

Every time I looked I was captivated by the view of something new and surprising that would present itself through the perpetually iridescent translucence of the Earth's colors against the blackest black, which seemed to go on to infinity. That view would always remind me, *Nikki, you're on a spaceship!*

Whenever I went to the window to catch a glimpse of Earth, I could never get "just a glimpse." I would inevitably be sucked into the vortex of the gorgeousness below. I found it necessary to set an alarm on my trusty OMEGA X33 (best alarm both on and off the planet—yes, that's an endorsement) to remind me to go back to work. Without the alarm, a ninety-minute full lap around the planet would go by before I'd notice any passage of time. With my eyes and mind and heart open to what I was experiencing, I was captivated every time. Enthralled by what I saw.

The planet looked alive.

It was like I was watching the Earth with the mute button on. The white swirly clouds of a hurricane moving across the ocean appeared soft and billowy, quiet and beautiful. I was startled by the thought of how different that hurricane must be for the people who were in the middle of it.

I became more familiar with the geography of the planet. I could tell which continent I was looking at not just by the shapes but by the colors and patterns. I noticed a tiny heart-shaped island in the Red Sea, a curve of the Amazon River that resembled the profile of an elephant, the trail of atolls in the Pacific that looked like footprints on the ocean, the patterns on the desert sands that looked like the tracks of chickens running all over them, and the dotted lights at night on the planet's surface that showed me where the people were and where they were not. I couldn't help but think that God must have a wonderful sense of humor, and that there is purpose in our discovery of new perspectives like this on our planet home. Although I was on a spaceship 250 miles above Earth, separated from all the people I know and love—and all the people I don't know too—I was presented with the stunning reality of our interconnectivity. I felt more connected to everyone and everything below me than I ever had before with my feet planted on Earth's surface.

Though I continued to want to see my home state of Florida, very quickly I shifted to a different perspective: Florida became a special place to see on my home *planet*. From space, I not only saw that Earth is a planet, but that this planet is my home. I was in awe and humbled.

Throughout each mission, what kept me grounded in space was the experience of looking out the window and feeling that connection to home. It was like an invisible wire connecting me to "ground," to Earth, to home.

In 2015, I grounded myself: I made the difficult decision to retire from NASA. My next flight assignment would not have been too far off in the grand scheme of flight assignments, so it was a decision to remove myself from any opportunity to fly in space again as a NASA astronaut.

What was I thinking? I had performed well in space. It felt almost natural to be there, yet my gut was telling me it was time to move on. I could not make this decision without a deeper inquiry, so I came up with a series of questions to consider.

1. *"Is it important for me to fly in space again?"* Seems like a simple question, but it isn't—especially knowing what an awesome and life-changing experience it is to fly in space. But was it important for *me* to fly in space again? I wrestled with this question for a bit until I was finally able to answer it in a simple and honest way. No. The answer was no. Big step!
2. *"Do I need to keep doing the work I'm doing in the astronaut office when I'm not flying in space?"* Ninety-nine percent of an astronaut's job is not flying in space. It's done on the ground. Of the fifteen years I was an astronaut, I spent only 104 days in space. The work astronauts do on the ground is critical to supporting the people and hardware that we send to space and to designing hardware for future missions. This question was a little easier to answer. No, I did not need to stay in the astronaut office to keep doing related work. I knew that even outside the astronaut office, I could still consult, advise, and participate in the program from the ground, providing value and having a positive influence on whatever would be going on.
3. *"Would I be okay letting go of all the other things that astronauts get to do here on Earth?"* I guess this question came from a more selfish standpoint, but I had to be comfortable with the fact that retiring from NASA would also mean that I'd give up a lot of cool, fun stuff that astronauts do here on Earth. I wouldn't be flying in T38 jets anymore; I wouldn't be diving in the Neutral Buoyancy Lab practicing spacewalks anymore; I wouldn't be controlling robotic arms or crawling around the ISS in beautiful VR simulators anymore; and I wouldn't be working every day side by side with a team I believe is made up of the most capable, talented, and extraordinarily competent people on the planet. While bummed not to have the opportunity to do

these things anymore, I knew I would still be in touch with the people, so the answer was yes.

Having taken a few months to consider and answer all these questions, I felt refreshed. Being an astronaut was a dream job, but I discovered that it should not, could not, be the final destination or the ultimate achievement in my life. I am thankful for and proud of the work I did. Being an astronaut was an extraordinary experience that opened up opportunities that might not have existed for me otherwise. With those advantages came the responsibility to pay it forward. It was time to embark on my next mission in life, and I knew that by "grounding" myself I wouldn't be stifling myself, but rather, I'd be lifting myself up and opening up to more wonderful opportunities.

When considering what these opportunities might be, I was certain of one thing: I needed to make sure that whatever my post-NASA mission was, it would allow me to share my spaceflight experience in a way that could benefit others. Because flying in space is impactful in so many ways, I think that finding a personal and meaningful way to share it is important to every astronaut. I kept coming back to the experience of painting in space—would art be my next path? I wondered how I could combine my artwork with my spaceflight experience and share it in a way that would engage audiences that might not even know there is a space station. I believe that everyone should know about the ISS and all the work there that's dedicated to improving life on Earth. And I want everyone to benefit from the three simple lessons we learn from spaceflight: we live on a planet, we are all Earthlings, and the only border that matters is the thin blue line of atmosphere that protects us all.

I was attracted to the idea that what I created through my art could help me do what so many of my colleagues have done: take my work as an astronaut and my time in space and bring the experience back to Earth to benefit humanity. Through my paintings, my intent is to share what I saw and felt through the spaceship windows. Nothing else in my life has offered me a more heightened appreciation for our relationship with our planet and one another. I'm hopeful that my art will encourage everyone to find their own way to experience that connection.

Coming back to Earth, I also searched for a way to have an experience similar to looking out the window of my spaceship. The two closest terrestrial experiences I've found are meditation (if you ever get a chance to try it, meditation in a float tank takes the experience up another notch) and something called "Earthing." I'd never meditated or practiced Earthing before going to space. I do both now. Both are practices of mindfulness and transcendence that connect what we feel and experience through all five of our senses to what we feel and experience with our mind and our heart. Meditating and Earthing are like looking out those spaceship windows.

My introduction to meditation came shortly after I retired from NASA, when I was interviewed by a gentleman in Australia, Tom Cronin. He's the creator of the Stillness Project, a global movement to inspire one billion people to meditate daily. When we met, he was producing a documentary film called *The Portal*, which presents a powerful argument for why meditation is the answer to a better life and a better planet. The practice helps us bring peace and calm into our own lives, so that we can bring peace and calm to a stressed world. The film inspires viewers to carve out space for stillness in their lives—to stay grounded. The tagline for the film is "Calm your mind. Open your heart. Transform the world."

As a thank-you for the interview, Tom provided me with three guided lessons to help start my own meditation practice. Tom was in Australia and I was in Florida, so these lessons took place through our computer screens. I discovered that what I had been describing as sort of a transcendent experience in space was one of stillness and calm. Tom showed me how a practice of meditation here on Earth could help carry me back to space and to the still and healthful feelings I had while gazing at the beauty of our planet through the spaceship windows.

Earthing is something I discovered while searching the internet for a yoga mat. It is very simply the practice of taking off your shoes and walking barefoot on the planet. In becoming conscious of that connection between your feet and the Earth, you're reminded of your connection to the Earth (we live on a planet) and to all the life that surrounds you (we are all Earthlings). Interestingly, making a connection between your feet and Earth literally grounds your body, so I guess it's no coincidence that electrical grounding is also sometimes referred to as Earthing. Scientific research has revealed surprisingly positive and overlooked benefits of making a connection with the surface of the Earth, including better sleep, reduced inflammation and pain, and decreased stress levels.[2] Every day we should take the time to be present in our bodies and connect with the Earth, to look up from our place on the planet and appreciate the gift of the thin blue line that surrounds us.

Obviously, there was no way for me to plant my feet firmly on the ground or to walk barefoot in the grass while floating in space on the ISS, but I was able to experience an equally transcendent connection with Earth through the spaceship window. In doing so, I realize now, I was experiencing a form of Earthing. It wasn't a physical connection, but it was a connection unlike anything I'd ever felt before.

It was clear to me that even though I was on a spaceship circling the planet, I was still part of the planet—I was still at home. The calmness I felt allowed me to be present in my life and work on board the ISS and also to maintain a presence at home through communication with my family on Earth. To feel grounded this way while in space allowed me to stay relaxed, calm, and aware regardless of what was going on around me. It helped me to be a better crewmate.

Here on Earth, I've found that being grounded can bring the same benefits as in space. A first step is to realize that this sense of transcendence—of being grounded—comes from the realization that you are one small yet meaningful part of a greater whole, that your personal health and planetary health are connected, and that we are all Earthlings.

Meditation and Earthing are two ways of staying grounded that have worked for me. It's important for each of us to find and practice some approach that brings this kind of transcendence and balance in our lives. Some common benefits of staying grounded that I've discovered in my research on the subject (and that I've experienced in my own life) include establishing meaningful connections with people and the planet; appreciating, and taking action, based on both the beautiful and the not-so-beautiful realities that surround us; believing that there is a solution to every problem; being grateful for Earth's unique way of supporting life; shifting focus from ourselves to others; experiencing awe and wonder at the world around us; and finding a higher purpose in life. I know this is a lot, but all these benefits, whether on a spaceship or a planet, help put us in a position to be the best crew members possible. To be grounded opens up our hearts and minds to our own mission in life and helps us find our way to fulfilling it.

In January 2016, six months or so after retiring from NASA, I found myself spending a lot of time visiting hospitals—children's hospitals in particular. My friend Gordon "Gordo" Andrews—one of the most creative and thoughtful people I know—had contacted me about a space-themed art project that he was facilitating in his work with the ISS program's Strategic Communications Group. We arranged to meet for coffee in the cafeteria at NASA's Johnson Space Center, so he could share his idea with me. He invited me to combine my love of art and my love of space by helping him work with an artist on a space-themed art therapy project at one of the local children's hospitals.

Something about walking into a hospital always gives me a churning, uneasy feeling in my stomach. I become hypersensitive to the smells, the lighting, and the plastic sterility of an environment filled with people who don't want to be there any more than I do. When I'm in a hospital as a visitor, that churning feeling is a mix of fear, anger, helplessness, and worry that someone I care about has to be there. But I also feel hope: maybe there's something I can do to bring comfort to, or help the recovery of, the person I'm there to visit. Because of all these "feelings," I was both unsure about Gordo's project and excited to learn more about it.

Gordo explained that he'd been contacted by a gentleman named Ian Cion, an artist who was the founder and director of the Arts in Medicine Program at the University of Texas MD Anderson Children's Cancer Hospital in Houston. He had approached Gordo with the idea to create a large-scale, space-themed art project with the children at the hospital.

Ian had been working for several years with children in hospitals on what he called "community" art projects. The children understood that their individual piece of art would be part of a larger collective piece to be made available for public display. All of Ian's work focused on using art to help patients cope with the experience of cancer treatment, and he was especially interested in using art as a tool for reducing pain and fear.

Before approaching Gordo with the space art idea, Ian had brought children and art together through projects like wrapping city buses with art and creating murals on hospital walls, animations, and a large dragon called "Okoa the Wave Rider." Okoa was twenty feet long and nine feet tall, and each scale was created out of the individual and very colorful artwork of over 1,300 patients. When I asked him, "Why dragons?" Ian said, "Dragons are protectors. The dragon represents the idea that even in the face of grave danger, there is hope and happiness; that with faith, courage, and love, one can see the beauty and miracle of creation, and see that we are part of something larger soaring through the universe."

I have to hand it to Ian, who also is an aspiring astronaut. When he initially approached Gordo, he did so with the idea that the best way to publicly display whatever space art he created with the children would be for him to fly to the ISS with it—the ultimate NASA artist-in-residence program. To Ian's dismay—though I think he understood it was an astronomically long shot—that flight has not yet happened, but something else spectacular has happened instead.

Working with Ian and Gordo, we decided that creating spacesuits from the individual artworks of many different children would be the best project to start with. At the time it just seemed like the most obvious and cool symbol of astronauts and flying in space, but thinking about it now, I realize that spacesuits also are "protectors"—just like Okoa the dragon. Gordo arranged for Ian to meet with the spacesuit team at ILC Dover (the company that makes our astronauts' spacesuits) and learn all about how the suits are constructed, so he would have a better idea how to design the plan for the work with the kids at the hospital.

The folks at ILC Dover loved the idea of an art spacesuit. David Graziosi is a senior engineering fellow at ILC Dover who has helped design and develop the NASA spacesuits in use today on the ISS and those that will be used in the future as we go back to the Moon and on to Mars. David is the one who made it possible for ILC Dover to help us work on the construction of the art spacesuits.

Now we had a plan, and we could start working with the kids at the MD Anderson Children's Cancer Hospital. This was the beginning of what we called the Spacesuit Art Project. Ian, Gordo, David, and I spent time at the hospital painting with the kids on small squares of fabric. We collected enough pieces of the children's art for a spacesuit we named "Hope." What began with the artwork from the children in this one hospital for one art spacesuit has since grown to five art spacesuits created with the artwork of children in hospitals and refugee centers in over fifty countries. The five spacesuits are called "Hope," "Courage," "Unity," "Victory," and "Exploration"—all named in honor of the strength and vision of the children who helped create them.

The Courage, Unity, and Victory art spacesuits have each been flown to and from the ISS. While the suits were in space, we held special video conferences between the children, mission control, and the crew on board the space station, which gave the children the opportunity to see their art in space and to speak with the crew in space wearing the suits.

As the Spacesuit Art Project grew to a global scale, we formed the Space for Art Foundation, which has allowed us to engage with hospitals and refugee centers and children all over the world. The four founding directors include Ian, David, Maria Lanas, and me. As a NASA employee, Gordo wasn't able to officially join the foundation, but he continues to be our ambassador with the ISS program office. None of this work would have been possible without him.

Maria Lanas is an artist who created an art and cultural exchange project called Projekt Postcard "One World," which connected kindergarten students in Croatia, Ecuador, Iran, and the United States through artwork and video conferences. The result of this project was not only new friendships between the children but also a beautiful collection of the children's art that was on display for three years (2016–2018) at Dulles International Airport in Virginia. In 2016, I was at Dulles on my way to meet Ian for the start of our "Unity" world tour—a trip he and I took to paint with kids in our ISS partner countries—when I was captivated by Maria's Projekt Postcard display. Through the magic of social media, I got to meet Maria, and we worked together to create the Postcards to Space Project, with artwork from over a thousand children in ten countries. Maria has been an integral part of our Space for Art crew ever since.

In addition to our four founders, we are fortunate to have the continued and very generous support of ILC Dover, which continues to help us turn the children's art into art spacesuits; David Delassus and his team at ABlok Interactive Experiences in Germany, who create online interactive experiences from the children's artwork (like the "Postcards to Space" art video that was sent to the ISS); and Unity Movement Foundation, which was formed by our dear friend Alena Kuzmenko in Moscow to spread the healing power of art and the inspiration of space to children and adults in cancer treatment all over Russia. Alena has participated in one way or another in all our art projects around the world.

The Space for Art Foundation grew from the work of the Spacesuit Art Project. We are on a mission to unite a planetary community of children through the awe and wonder of space exploration and the healing power of art. Our strategy is based on continuing to facilitate and implement space-themed art and healing programs, awarding scholarships and grants to art students and artists who want to develop their own space-themed art therapy ideas, and documenting and sharing our work in support of ongoing research in the field of art and healing. At the time of this writing, we have created eight planetary community art projects with over three thousand children from hospitals, refugee centers, and schools in over fifty countries—and for our current spacesuit art project, named "Beyond," we are hoping for the participation of children from every country on the planet. When the art spacesuit is completed, it will be put on display at the UN Climate Change Conference (COP26) in Glasgow, Scotland, in November 2021. And maybe (fingers crossed) it will someday travel to the ISS.

Since that first meeting with Gordo, I've participated in many art sessions with children with cancer, who are going through what you hope is the worst thing they'll ever have to deal with in their lives. It still does not make sense to me that these three words—"children with cancer"—should ever have to be said together, but sadly, they are. The World Health Organization reports that cancer is a leading cause of death for children and adolescents around the world; approximately 300,000 children ages zero to nineteen are diagnosed with cancer each year. In the United States, cancer is the number one cause of children's death by disease.[3] The American Cancer Society reports that approximately 11,050 children in the United States under the age of fifteen (and almost 6,000 more between ages sixteen and nineteen) will be diagnosed with cancer in 2020—that's roughly 45 children in the United States per day who are diagnosed with cancer. Thankfully, treatment has improved, and the overall survival rate for childhood cancer has increased drastically over time, from roughly 10 percent survival in the 1970s to over 80 percent in the 2010s. However, while the overall survival rate has improved, the survival for many rare childhood cancers continues to be far less. Further, childhood cancer diagnosis rates have been rising slightly over these same few decades. Moreover, more than 95 percent of childhood cancer survivors are impacted for the rest of their lives by significant and chronic health issues related to the long-term toxicity of the cancer treatment itself.[4]

While the causes of childhood cancer are not completely understood, scientific research has not only demonstrated links between environmental exposures and the development of childhood cancers but also shown that childhood cancer survivors are more vulnerable to the adverse effects of environmental risks such as air pollution. Dr. Catherine Metayer, an adjunct professor at the University of California–Berkeley School of Public Health, along with her team, won a $6 million grant in 2016 from the National Institute of Environmental Health Sciences to study the causes of leukemia in children.

Dr. Metayer's research has shown that the upward trend in the diagnosis rates "points to the important role of the 'environment' in a broad sense, whether acting alone or in concert with genetic factors." "In the environment," she has noted, "a lot of things have changed. A lot of chemicals have been brought in. We are all exposed to many of them.... Research has tied a child's risk for leukemia to their exposure to pesticides and paint, to their father's smoking history, and to both parents' exposure to chemicals on the job." In her work, Dr. Metayer highlights "the need for prevention programs to reduce harmful environmental exposures and promote healthy lifestyles, especially targeting vulnerable populations."[5]

Cancer in children and teenagers is still considered uncommon and accounts for less than 1 percent of all cancer cases in the United States. Perhaps this is why only 4 percent of the billions of dollars spent in the United States annually on cancer research is directed toward treating children with cancer.

Many cancer-related studies have been performed on the ISS (including those referenced in chapter 1 by ISS astronaut Dr. Serena Auñón-Chancellor). The microgravity environment on the space station has provided researchers with new models for studying the disease and new formulations for therapeutics. For example, the way surface tension behaves in microgravity has allowed scientists to further refine a therapeutic technique called microencapsulation for improving the targeted delivery of cancer drugs. This targeted delivery of chemotherapy and radiation drugs reduces the negative impacts on the body caused by their toxicity.[6] This is an example of space research that will benefit children undergoing cancer treatment, but to my surprise, none of the cancer research to date on the ISS has been targeted at childhood cancer specifically. We are hopeful that our work at the Space for Art Foundation will help change this.

As I became more and more aware of what these children and families I paint with are going through, I found myself always asking, *Why?* The statistics are shocking, and the response to them is shocking. This is about children. I struggle to understand how any number of children suffering from cancer or any disease, or from any other threat to their happy and healthy lives, could ever be considered normal. I started to think about this more and more from the planetary health versus personal health standpoint, from the view of everything as interconnected. What we do to our planet, we also are doing to ourselves, and more importantly, to our children.

When I arrived at the MD Anderson Center Children's Cancer Hospital in Houston for one of our first art sessions, it was no surprise to me that I had that churning feeling once again, but this time it wasn't because I was visiting an old friend or a loved one in the hospital. I was there to facilitate one of our art spacesuit activities. We were at the hospital that day to work with the children to create colorful paintings on fabric that would be quilted together with the paintings of other children around the world into the "Unity" art spacesuit.

I go into these sessions praying that the work of the Space for Art Foundation will have a positive impact on these children's lives, and that the art and inspiration of spaceflight that we share with them will bring them some joy. I go in knowing there's no way for me to truly understand what the children and their families are going through, and I go in with a prayer of gratitude for the health of my son.

On that day, there were about fifteen young children sitting around a table. All of them were undergoing treatment for cancer. Some of their parents also were present, as well as hospital staff, a few of my astronaut and NASA colleagues, and Ian, David, and Gordo.

I randomly sat down next to a small eight-year-old girl who'd lost all her hair to chemo and wore a brightly colored hat. This insightful young lady and I began talking together about living in space. She told me about seeing stars, and we were pointing up and imagining the outer space that surrounds us.

Then, as she applied paint to her brush and dragged it along the fabric, she casually said, "Being an astronaut must be a lot like what happens to me."

I was startled by her words, but kept smiling and asked her to tell me what she meant.

We kept painting, and after a while she matter-of-factly compared being an astronaut to experiencing treatment in the hospital.

"You don't get to see your mommy and daddy and friends the same way, you don't get to go outside anytime you want, you have to eat different food, your body's changing, they do all kinds of tests on you, and I think you have the radiation too."

Holy moly.

She then shifted to telling me that she liked to paint at home. I told her about painting in space, and she thought that was cool and said she hoped that maybe one day she'd be able to paint in space too.

Ian, who is the artistic genius behind these art spacesuits, had told me I would hear these kinds of "wise beyond their years" philosophical utterances during these visits. But never before had I heard anything so thorough and thoughtful from such a young person.

Every time I walk into a hospital wearing my blue astronaut suit, the hospital staff thanks me and says how inspirational it will be for the children. I am thankful that my presence as an astronaut might help them because that's what we're trying to make happen for these kids. But the truth is that every time I walk away from these events I'm left with the feeling that there's no possible way that any of these kids could be as inspired by me as I am by them, and by the hope, courage, and strength that I see in each of them and their families.

Each time I do one of these events, I am in awe of all the children. I see so clearly that through this intersection of art and space, there is healing. There's always at least one child at each event who obviously showed up only because their parents wanted them to. They slink into the room looking as if they would much rather go back to their hospital bed and sleep. At first, when they sit at the table, I can tell they aren't all that interested in painting. But then, after a few minutes, they sit up straighter, they chat with whoever's sitting next to them, they begin painting, they become excited thinking about space exploration, and they share stories of their own experiences. I see these children transcend their immediate circumstances and find the strength to not only sit up straighter and paint but also to imagine their own future outside the hospital.

Reflecting on the profound words of that little eight-year-old girl, I realize that she touched my heart. It was a moment of transcendence for me—I had found my "place." This was the day I discovered my next mission in life. Her words made me realize that I'd been blessed with the opportunity to fly in space so that I could come back to Earth and work on these projects with these children—bringing the inspiration of spaceflight and the healing power of art together for kids around the world. And it makes perfect sense to me now: we should always be looking for ways to use our experiences to help make life better—for others and ourselves.

Consider a crew on a mission to Mars. With our current propulsion technology, it would be at minimum a nine-month journey in a relatively confined spaceship just to get to the red planet, then at least another five hundred days on Mars to wait for the planets to align again properly for the first opportunity to return to Earth, and then the travel time of about nine months for the return. For such a trip you're going to want both a crew and a spaceship that are grounded.

My young friend in the hospital was correct when she compared her experience with cancer treatment to mine as an astronaut—you aren't able to see or speak to your family and friends the same way, you can't go outside anytime you want, your food is very different, your body is changing, lots of tests are done on you, and there's radiation. And like a child being treated for cancer who cannot know if her life will ever return to normal, at some point into the more than thirty-five-million-mile flight to Mars, astronauts, for the first time, will no longer see Earth recognizable as Earth out the window because it will have become just a dim dot of light.[7] (For reference, the farthest we've traveled as humans in space is to the Moon. While it's not easy to get there either, and it's a quarter-million miles away, it's like a quick errand to the corner store in comparison to the trip to Mars.)

It became clear to me that we will need to do something so that astronauts feel that grounded connection to Earth no matter how far away they travel—and to feel grounded in their new home. We will need activities similar to the hospital painting sessions with the child patients because the astronauts won't have the glowing, colorful view of Earth out the window anymore. To stay grounded, they'll need to be able to paint or play music or engage in activities that allow them to keep the human in human spaceflight. Whether it's painting or playing music on an iPad—or even better, the *Star Trek* holodeck—they'll need to be able to bring a piece of home with them on the journey, however small it is.

On my ISS missions I was able to bring some small items—a little bit of "who I am" on Earth—to space with me. I decorated my crew compartment with two little stuffed dogs from my son and pictures of family and friends, I had my favorite book to read, and I wore a T-shirt from my high school and my husband's wedding ring on a necklace. All these items gave me a sense of home on the space station and helped me share the experience with the people I care about back on the planet. Anything I chose to bring with me to space was important enough that I would want to bring it back to Earth with me too. As humans we have an intimate connection to our home planet. We are Earthlings. My favorite way of thinking of being grounded in space is that we are an extension of the Earth when we fly there—that we come from the Earth.

As we continue to travel farther from home, this sense of connection will become more necessary than it already is for low Earth orbit spaceflight. I'm sure that we are capable of making these voyages, but we should be thoughtful about accommodating not only the needs of the spacecraft traveling through such an extreme environment but also the emotional and psychological needs—ultimately the health—of the crew. The human science associated with these missions may well be more challenging than the rocket science. It will be increasingly important for astronauts to be able to work well as a team, to feel that their work is meaningful, and to be flexible in their approach to resolving problems and empathizing with their crewmates. Distances never traveled by humans before, radiation, confined space, communication delays of up to forty minutes, and all the other factors already so eloquently pointed out by our young cancer patients painting spacesuits will be part of this epic journey.

In the late 1980s, researchers began to focus on the salutogenic (or growth-enhancing) aspects of space travel. One thing we do know from the human spaceflight research to date is that all astronauts questioned reported positive changes as a result of flying in space; generally they described having a new appreciation for the life we share our planet with and for the planet as our one shared home. In 1987, my dear friend Frank White, a space philosopher and author, coined the term "Overview Effect," which is now commonly used to describe astronauts' experience as a result of seeing Earth from space. Frank formulated the Overview Effect philosophy when he was studying what people in the future might experience when they're living permanently off Earth. He is recognized as the leading expert in this field, and he continues to pursue his research on how the Overview Effect can help us, as humans, fulfill our larger purpose.[8]

Frank first considered the idea of the Overview Effect when he was off Earth himself. Flying across the country and looking out the window of an airplane, he was imagining what it would be like to live permanently off the Earth—on a space station somewhere between the Earth and the Moon, a vantage point that would still provide a view of Earth. From his view thirty thousand feet above the Earth, all of what he saw below appeared small and insignificant.

"I knew that people down there were making life and death decisions on my behalf and taking themselves very seriously as they did so," he told me. "From high in the jet stream, it seemed absurd that they could have an impact on my life. It was like ants making laws for humans.... I knew, though, that when the plane landed, everyone on it—me included—would act just like the people over whom we flew." He continued with the thought: "If I lived [permanently in space], I would always have an overview of the Earth. I would see it from a distance. And I would see it's a unified whole. There are no borders or boundaries. All of these realizations would become knowledge. Which, living on the surface, we find it very hard to philosophically grasp, or mentally grasp. And the term overview effect came to me....

"People who live in space will take for granted philosophical insights that have taken those on Earth thousands of years to formulate."

Frank formed his Overview Effect hypothesis that our perspective changes when we see the Earth from space—when we have "the experience of seeing firsthand the reality that the Earth is in space"—after his personal experience on the airplane and from his musings about what permanent settlers off-planet might experience. To investigate further, he turned to the closest people he could find to off-planet settlers—astronauts. Frank's research and that of other human spaceflight researchers have shown that after returning from space, astronauts report higher levels of universalism—that is, a greater appreciation for other people and nature. This was certainly my experience, as I've tried to convey through my own Earthrise moment and expression of the three simple truths: that we all live on a planet, as Earthlings, protected by the same thin blue line.[9]

Researchers recognize the psychological benefits that astronauts experience through Earth-gazing, which include "reduc[ing] stress and even inspir[ing] spiritual or transcendental experiences," according to Kirsten Weir in her 2018 article "Mission to Mars," citing Dr. Nick Kanas's review of the psychosocial issues related to long-distance space travel. Kanas cautions that such benefits "won't be available from 35 million miles away. Nobody knows the effect of seeing the Earth as a dot in the heavens. Maybe it won't have any effect—but maybe it will."[10]

Because "maybe it will," NASA and the other international space agencies continue their research into the psychological challenges of travel into deep space, so as to address the human needs associated with this type of journey. I for one believe that there will be a shift in our human response to the spaceflight experience when we can no longer make that visual connection to our planet.

Some of this research looks at utilizing tools and countermeasures to relieve stress and support mood enhancement and restoration, such as virtual reality (VR), augmented reality (AR), digital therapeutics, and even brain stimulation.[11] These tools have some things in common, including incorporation of nature experiences, sensory immersion, and ways of sustaining a connection to Earth. The NASA Human Research Program has contracted with the Baylor College of Medicine's Center for Space Medicine to help solve the challenges of human deep space exploration through a program called Translational Research Institute for Space Health (TRISH). TRISH is working to find the latest and greatest innovations in science and medical technology to help NASA put humans on Mars while also adapting these innovations to improve life on Earth. Their charter is to push the limits of technology by investigating the viability of ideas that seem way out of reach at this point.

One study that I find interesting is being conducted in Antarctica, the most otherworldly and remote place on Earth, and is led by a retired astronaut colleague, Dr. Jay Buckey. Jay flew on Space Shuttle *Columbia* mission STS-90 in 1998, which was a sixteen-day life sciences mission called Neurolab. The astronauts performed experiments with a focus on the effects of microgravity on the human brain and nervous system.

Since 2001, Jay has been a professor of medicine at Dartmouth College's Geisel School of Medicine, where he is also director of the Space Medicine Innovations Laboratory. One of his areas of study has been testing and developing self-directed, autonomous, interactive, media-based behavioral health tools and virtual reality for use in isolated and confined environments (for example, Antarctic research stations and space stations). He and his colleagues are evaluating the use of this technology to help people in confinement deal with three main issues: conflict resolution, stress management, and mood improvement.

Jay was asked in a 2020 interview about how people were dealing with isolation during the COVID-19 pandemic. "Living in isolation and confinement is challenging and it's challenging for everyone," he confirmed. "Long-term relationships matter, it's important for us to be able to deal with conflict, to deal with stress, to maintain our mood, and also doing it in a way that we are maintaining our relationships that we value."

He added: "In the space program or in the Antarctic station, you are depending on your crew mates and the people you are with for your existence. We can certainly see that in a space environment or in Antarctica, but in our daily lives it's true, too."[12]

Virtual reality forests, beaches, and cities are being tested at Australia's Antarctic stations to see whether they provide support for expeditioners there; this research will also inform the development of programs supporting astronauts on long-duration space flights, like a mission to Mars. In a 2018 interview, Jay explained that VR, by allowing people to virtually immerse themselves in natural settings—"they can be in the Bavarian Alps, or they can be on a beach in Australia"—has restorative effects similar to the effects of exposure to nature itself: "it can help people to relieve stress [and] it can also help perhaps improve people's attention and mental functioning."[13]

Many research studies have sought to better understand the impact of our connection to nature on our health and well-being, and several concepts and theories have resulted from them. For example, the biological theorist E. O. Wilson's "biophilia" concept, as mentioned in a recent *Washington Post* article, is the hypothesis that "over millions of years humans developed evolutionary preferences for lush, abundant environments because they offer a space to recover from stress and fatigue." In a similar vein, the "attention restoration" theory developed in the 1980s by Rachel and Stephen Kaplan, both University of Michigan professors known for their research on the effect of nature on people's relationships and health, "suggests that the mesmerizing quality of nature... allows us to recuperate from attention-sapping and often hyperstimulating modern life."[14]

Other studies, like the one in Antarctica, have pushed beyond trying to understand the impacts of real nature to observing how humans respond to "nature" through VR. One of these studies was done in 2020 by psychologists at the University of Exeter and the University of Surrey in the United Kingdom. They found, according to one coauthor, Dr. Patrick White, that "virtual reality could help us boost the well-being of people who can't readily access the natural world, such as those in the hospital or long-term care. But it might also help encourage a deeper connection to nature in healthy populations, a mechanism which can foster more pro-environmental behaviors and prompt people to protect and preserve nature in the real world."[15] For our health and well-being here on Earth, and even as we venture out to establish our presence in a distant place like Mars, we will still be searching for (and finding) more ways to ground ourselves as Earthlings.

In 2015, on the same day I retired from NASA, I flew to New York City for the World Science Festival. It was there that I had my first introduction to The Studio, a department at NASA's Jet Propulsion Laboratory (JPL) in Pasadena, California.[16]

I arrived at Gould Plaza on the campus of New York University, where I was scheduled to make a presentation in the NASA Orbit Pavilion about photographing Earth from space. I had no idea what an Orbit Pavilion was, but I imagined it might be a large tent set up by NASA to accommodate presentations. When I saw it, I was happy to discover that it was not a tent, but a large aluminum spiral structure that reminded me of a nautilus seashell laid on its side. This massive, interactive art installation was created by the team at The Studio at JPL in response to their own question: "What if you could make contact with the nineteen orbiting Earth science satellites that are passing through space continuously to study the planet, collecting data on everything from hurricanes to the effects of drought, and bring them 'down to Earth'?"

The Studio's answer was to tie sound to the trajectories of the satellites as they orbit the Earth, allowing us to "listen" to their movement. Each time a satellite passes overhead, it triggers its own soundtrack within the spiral, which allows you to "hear" the location of the satellite. As the satellite moves through space, the sound, which might be a crashing wave to represent one satellite or a frog croaking for another, also moves across the twenty-eight speakers in the pavilion walls. The Studio team brought these satellites "down to Earth" through this structure and soundscape for all of us to experience. That day I had the honor of presenting photos I'd taken of our planet from space inside this massive, structural piece of art. I was so impressed by the Orbit Pavilion that I did some research about The Studio and about those who worked there, and that's how I came across Dan Goods, who is The Studio's founder and its lead "visual strategist."

"I am passionate about creating moments in people's lives where they are reminded of the gift and privilege of being alive," Dan said during our interview, which I conducted by video conference from my home "office" right off my back patio while Dan sat in a lawn chair in his own backyard. By experiencing the variety of sounds of nature that are "coming" from the satellites, as well as the light that shines through the pavilion structure's patchwork of metal, Dan explained, "we want people to have a connection with this network of satellites orbiting above us that we can't see, and for this to be a space that's contemplative. It's an object of wonder. You go in and maybe close your eyes, and concentrate on the location of sounds. When you're looking at too much stuff, you don't focus on listening as much. It's been beautiful to walk in there and see people with their eyes closed."

On his website, Dan describes himself as "a curious human, an artist, creative director, and public speaker...a creative problem solver...looking for that which stirs the mind and soul." Perhaps this is why every time I see him he looks like he's deep in thought, perhaps imagining his next project.

After graduating from the ArtCenter College of Design in Pasadena, California, Dan had a desire to apply his design skills to science.

He met Dr. Charles Elachi, who, as the director of the JPL at the time, was seeking ways to better communicate JPL's complex work to the public. The JPL is a NASA center that, year after year, has been recognized for working at the edge of what's possible in space exploration. JPL is responsible for robotic exploration of the solar system and operates nineteen spacecraft and ten major instruments that are carrying out planetary science projects, Earth science, and space-based astronomy missions, including the rovers on Mars.

"I had about two seconds to sell myself," Dan recalled, "and I said, 'Wouldn't it be cool to have artists help come up with new ways of space exploration?'

"He said to me, 'I don't really understand what it is that you do, but I'll give you six months.'"

That was in 2003. Since then, Dan has grown his team to five very talented visual strategists (these are creators of visual strategies for communicating with different audiences), including David Delgado, the person he has referred to as his "creative soul mate" since the days when they were classmates at the ArtCenter.

When you hear about rovers like *Perseverance* on Mars, or spacecraft like *Juno* at Jupiter, *Cassini* at Saturn, *New Horizons* at Pluto, or the Voyager program studying interstellar space or discovery of exoplanets, that's JPL. Dan's team of five at The Studio is made up of multitalented people who are not just artists and designers but also strategists and thinkers who are passionate about working with scientists and engineers to imagine the future and solutions to complex problems by helping them "think through their thinking." They also help the scientists and engineers connect with the public by giving people unique experiences that help them understand the work being performed by JPL, but, more importantly, by leaving them with a sense of awe about the universe.

While a number of The Studio's projects are large-scale installations like the Orbit Pavilion, one of the most impressive is its smallest, though you wouldn't guess its size by its name. It's called the Big Playground and consists of six rooms filled with big piles of sand and one very special display: a single grain of sand with a tiny hole drilled into it.

The grain of sand represents our Milky Way galaxy. The tiny hole was drilled into it to show not only where we live in the Milky Way, but also where we have already found thousands of other planets around other stars within that tiny portion of our galaxy.

Now get this: if the grain of sand represents our entire galaxy, then you would need those six *rooms* of sand to contain all the galaxies in the known universe. Through this creative display, The Studio has given us the opportunity to see the universe through a hole drilled into a grain of sand. (I have to admit that I still haven't gotten over the idea of *a hole drilled into a grain of sand!*)

I shared with Dan the impression the Big Playground had made on me. "It gave me goose bumps," I told him. "At the same time it brought me back to a vivid childhood memory I have of 'experiencing' our universe through a pullout poster in a *Nat Geo* magazine, and it confirmed again for me the significance of our tiny place in the ginormous universe."

I think often of that childhood experience. I must have been about ten years old. My parents, like many back then, subscribed to *National Geographic*. (Our bookshelves also held sets of encyclopedias—the olden day's internet. I wish I'd saved them.)

I shared the scene with Dan.

"I sat on my bedroom floor with the latest issue and spread out to the centerfold. I always went to that place in the magazine first because there usually was a cool poster. This issue was about space, and yes, the poster was cool. It was an image of a black and colorful elliptical-shaped form covering most of the poster's white background. Contained within that ellipse was our known universe. When I think about it now, that concept of the known universe is still incomprehensible to me (maybe because it's even more difficult to fathom the *unknown* universe).

"At age ten, looking at this poster, I was first captivated by its beauty. I looked for the 'you are here' place in the picture. I next shifted to staring at the picture with questions: The idea of a *known* universe seemed reasonable, but what was all the white stuff around it? The unknown universe? Heaven? Infinity? And then my gaze moved from the edge of the white paper to the expanse of my bedroom floor and out my bedroom window."

Dan was smiling big as I shared this experience with him because it's that kind of visceral "being left with a lasting memory" feeling that the team at The Studio wants everyone to have.

"One of my favorite moments at JPL," he told me, "was when I showed the grain of sand with the hole drilled into it (under a microscope) to a guy who gets to point Hubble [*the* Hubble Space Telescope] at places in space. 'I get to tell it to point that way,' he told me. LOL. So he looked at the hole drilled into the grain of sand and he looked up at me and stared for a couple seconds and then said, 'You reminded me why I work here.'

"I was like, that's amazing! As an outsider, you forget that when you're in the midst of all the details and all the bureaucracy, you kinda forget the big, cool, emotional reason why you wanted to be there in the first place. That was a really beautiful moment."

I think of all the planetary exploration that's going on with JPL's robotic spacecraft, taking us farther and farther away from Earth and bringing all its discoveries back to us, forming a connection between us and the universe that surrounds us. One beautiful example of many is the *Cassini-Huygens* mission to Saturn. After two decades in space, it reached the end of its remarkable journey in 2017, having provided invaluable scientific information and imagery of Saturn and its many moons. What it also did was expand our understanding of Earth's relationship with this gaseous and iconic ringed planet—one particularly stunning image shows Saturn and its rings with a tiny dot of light below it. That dot of light is us.

One of the missions of the entire JPL team, including The Studio, is to discover more planets around other stars, and they are finding some that might even support life like ours here on Earth. You might think that the study of such distant places would leave the researchers feeling unsure about who and where we all are together here in space, but it doesn't. Dan and every other person I've spoken to who is involved in this type of exploration and research has told me just the opposite—that by understanding more about the universe, we understand ourselves even better.

Bringing Dan's story back to Earth again, he shared with me his Earthrise moment.

"It was like a bolt of lightning that hit me, and I'll remember it the rest of my life," Dan said. "It happened as my friend and pastor Irwin McManus was speaking about how it's a gift and privilege to be alive. I had been thinking about work or something as he spoke, and I remember how what he was saying somehow came through to me. I really heard it.

"It is a gift and privilege to be alive! What am I gonna do with that knowledge? It's continually challenging me. It's continually making me appreciate what's going on around me."

Dan panned his camera around so I could see where he was in his yard, and he paused on one special place with a patch of long, delicate grass. "Right now I can just sort of see the grass blowing, and I love watching the grass blow. That's actually why we have that grass over there. My wife is amazing, and she's figured everything out about here [the yard], but I said, 'Can we just have some grass that blows in the wind? I need some wavy grass.' So she put that out here for me."

He described how looking at the wavy grass and appreciating its presence reminds him of the privilege of being alive, of the need to be intentional, of the need to be present and to ask himself, "Am I using my life meaningfully?"

"When I think about the meaning of the word 'grounded,'" Dan said, "I think it is to have a sense of humbleness. It means that you don't take things for granted. It means you understand the gift and privilege of being alive. To understand that everything doesn't revolve around me and to empathize and listen to others and make hopefully good decisions that think about humans in the midst of it. I think being grounded is that."

Dan's work at The Studio is meaningful. It is meant to draw attention to the awe and wonder of our universe and to help us embrace the awe and wonder in our everyday lives. By doing so, we ground ourselves in our own purpose and connection to the world around us.

Going to space brought me back to Earth. I realize this is kind of a strange statement, but it's true. I went to space to support the international mission of science and exploration, yet while I was there my connection to home, and my view of Earth as my home, deepened and expanded. When I was the farthest from the planet, I found myself feeling more connected than I had ever felt before when I'd had my feet on the ground.

This awareness of a planet, and all that goes along with it, as our home is both humbling and awesome. It's what kept me grounded while I was in space, and it has continued to provide a deep sense of groundedness now that I've come back to Earth more appreciative for and understanding of who and where we all are in space together.

As it has done for me, staying grounded can allow us to acknowledge reality in the simplest possible way, so we can respond in the most powerful way. That reality includes the three simple lessons we learn in space—we live on a planet, we are all Earthlings, and the only border that matters is the thin blue line of atmosphere that protects us all. The quality of life on Earth—the quality of our shared future—depends on the quality of our stewardship of the planet. When we aren't grounded, we are vulnerable and can be thrown off balance. The same is true for a spaceship and for a planet. Just as it's important that the ISS be grounded, and that we be good stewards of the artificial life support systems we've created to live a safe and healthy life there, so should we be good stewards of the life support system that has sustained us naturally for millennia on Spaceship Earth.

It shouldn't be a surprise that how we treat one another will affect the quality of our relationships with the people close to us, as well as with those "far away" who we don't even know. When we treat one another with consideration and kindness, generally our relationships are healthy and have a chance to flourish. The same is true for how we treat our planet. We need to establish a positive and nurturing relationship with our planet, just as we need to do so with all the life it supports.

Once again, I am reminded of the wise words of my young friend from the hospital. The life she described in the hospital is exactly what we can expect to experience in a not-so-far-off future if we fail to care for our environment today—living in isolation, having to undergo treatments and tests in a highly controlled and sterile environment. (At the time of this writing, we were a year into a global pandemic that gave us a glimpse of what a future like that might be like—but our COVID-19 experience has been gentle on us compared to what we would experience if our environment as a whole were compromised.) Just to survive on Earth we could end up with little choice but to live as we would on Mars. The state of our environment affects the way we live. How well we take care of one another and our life support system will determine how and how often we can have contact with our family and friends, the kind of food that will be available to eat, and the way our bodies will change in response.

There's no denying the underlying complexity of these ideas when we think about them on a planetary scale and begin once again to feel overwhelmed when we ask: What can I do? But the challenges can become less daunting when we connect these simple lessons to our everyday lives and remember that little things matter.

It has been found that people who regularly walk barefoot on the Earth and open themselves up to the good in life report higher levels of happiness and good health. The science is quite simple: There is an undeniable connection between planetary health and personal health. It is important for us to engage in a way of life that acknowledges our connection to the world and to remember that we live on a living planet.

Being grounded can help us solve planetary challenges. As we deal with our world's increasing complexity and uncertainty, mindfulness can help lead us to the solutions for our seemingly impossible problems. Find your way by keeping your mind and your heart open.

We must be mindful and thoughtfully invest in the care of our planet's life support systems. By doing so, we are caring for all life here. This is our best way to ensure that we leave this planet our children will inherit in a condition that will allow them to thrive (and give them the opportunity to care for it for their own children to inherit as well, and so on). What better way to show our love for them and future generations?

This mindfulness, this being grounded, is a call to action.

The awe and wonder of experiencing Earth from space is humbling, and it left me with a feeling of obligation to share it in a meaningful way. I've chosen to do that through my artwork and the work of the Space for Art Foundation, through this book, through my daily choices and decisions, through some level of activism, and by becoming more active in my local community while taking into consideration my planetary community. It is essential for all of us to commit to the changes and practices in our daily lives and in our communities that will add up to a meaningful difference.

Before my first spaceflight, I had never thought much about the fact that we live on a planet. I do now, every day. I'd never thought about my feet making contact with the Earth and what that means. I do now. I want everyone to have this feeling and to know that you don't need to travel on a spaceship to feel it. The same awe and wonder that comes from spaceflight and seeing Earth from space can inspire each of us here *on* the planet too.

One of my friends, and one of the most thoughtful people I know, Christina Rasmussen (author of the books *Second Firsts* and *Where Did You Go?*), invited me to be her guest on the first episode of her *Dear Life* podcast, where we discussed the transformative nature of my journey in space. Since then, I smile when I read her reflections on our conversation. She recently posted a beautiful picture of sunrise from her backyard and wrote on her Instagram page:

"I can only imagine what it must feel like to experience 16 sunrises in a day. The International Space Station completes one trip around the globe every 92 minutes, the astronauts experience 15 or

16 sunrises every day. Maybe we should pause every 92 minutes here on Earth too. To witness ourselves. Have a cup of coffee or tea. And reflect. Just like we do in the morning with our one and only sunrise. Here's to a Wednesday with a reflective pause wherever we can find it."

I strongly second Christina's suggestion—that we pause and reflect, whenever we can, on our place and shared lives and future together here on Earth.

Next time you're outside—whether looking across the water and out to the horizon at the beach or looking up at the stars on a very dark night in the middle of a park or forest (maybe watching the dot of light of the ISS flying over) or even in the middle of the city—take a moment to notice where you are standing. You are on a planet, spinning in outer space. If you can see the Moon or the Sun (or both at once) from where you are, recognize that you are seeing two or three heavenly bodies; you just happen to be standing on one of them. Take this moment to appreciate how incredible it is that you and I get to live on a spot in the universe where our every need for survival is provided. Ground yourself in that awareness. Use it in your life.

This stunning silhouette of the Space Shuttle *Atlantis* was taken from the International Space Station, capturing my ride home after my first mission on the ISS.

NASA

CHAPTER 7

WHATEVER YOU DO, MAKE LIFE BETTER

AS I PREPARED TO LEAVE THE ISS, I felt a wistful gratitude, knowing that I might not ever be in space again, but also understanding how blessed I was to have been there at all. Knowing that I might not float and fly in microgravity again, but thankful that it had become natural to move about in three dimensions. Knowing that I might not have the opportunity again to see firsthand the glowing and colorful view of our planet in space from that very special vantage point, but equally sure that its impact would be part of me for the rest of my life.

After my three months on the ISS, I rode home on board the Space Shuttle *Atlantis*, with her six-member STS-129 crew made up of my friends Charlie Hobaugh, Butch Wilmore, Leland Melvin, Randy Bresnik, Mike Foreman, and Bobby Satcher. They floated into the ISS on November 18, 2009, and spent six days working with the six of us already on station. In addition to the busy work of the mission schedule, one of my favorite memories of all of us together was the celebration of my forty-seventh birthday (made extra special with their delivery of Blue Bell vanilla ice cream cups and homemade chocolate cake by my dear friend and fellow astronaut Marsha Ivins).

In the final days before returning home, I often tried to steal one more look out the window or one more rolling and somersaulting flight from one end of the station to the other. I'd touch my hands to the spaceship, trying one last time to capture the essence of this place, and the uniqueness of the experience that was now a part of me. With me all the time, though, was anticipation of the joy of reuniting with my family back home on Earth.

On the day of our departure, November 24, 2009, in the final minutes before I floated through the hatch, out of the space station, and into the Space Shuttle *Atlantis*, I gave each of my ISS crewmates a strong and long hug, and then we all had one final group hug.

As they released me and sent me floating toward the hatch, I felt a final smack on my back. Jeff Williams—who would be staying on board the ISS, along with my other ISS crewmates Max Suraev, Bob Thirsk, Frank DeWinne, and Roman Romanenko, to complete their missions—stuck a green cargo label on my shirt that identified me as BACK TO EARTH ITEM #914.

Once on *Atlantis*, we worked through the steps that allowed us to undock from the space station. As pilot Butch Wilmore pulled *Atlantis* away from the station, commander Charlie Hobaugh announced, "Houston and station, physical separation." The sadness came as I watched us move away. Over the radio, I heard the traditional ringing of the bell and Jeff Williams's report from the ISS to mission control, "United States Space Shuttle *Atlantis* crew and International Space Station crew member Nicole Stott departing."

My face was glued to the window throughout all of this as we flew a perfect loop around the station, and my gaze remained fixed on the ISS until we had separated so far apart that the space station had shrunk to a tiny dot of light and then disappeared from sight.

Two days later, as we prepared for our descent back to Earth, we went about getting dressed for the trip. Our orange launch and landing spacesuits, called advanced crew escape suits (ACES), are designed to protect astronauts in case of an emergency during launch or landing. Under the orange suit, we wear a blue cooling garment that is like thermal underwear with an extensive network of tubes woven into them; that allows us to plug into a portable cooler to pump cold water through the tubes. The orange suit itself is a pressure suit made of a rubber liner and a flame-resistant Nomex orange outer layer that is designed, when integrated with the gloves, boots, helmet, and parachute, to help keep an astronaut alive during emergency events inside the shuttle or if the astronaut needs to bail out. Each astronaut also carries a survival kit and radio in their suit's leg pockets.

The last time we'd worn these spacesuits was when we left Earth. The suits had been packed safely away since we'd taken them off shortly after our arrival in orbit. I chuckled when I remembered suiting up for launch. Even though we had been trained on how to put the suits on, it had required the assistance of a highly specialized team of people who were responsible for ensuring that all was well with our suits and that we were properly fitted into them. The team was so meticulous that it seemed they'd have preferred to get us into the suits without us having to touch them at all.

But in space, we were on our own. While we did our best to be diligent and deliberate in handling the suits and getting into them, to do this while floating had a free-for-all feeling about it. One of the oddest sensations was feeling like we were being "birthed" as we struggled to pop our heads through the rubber seal around the neck ring.

As we got into our suits, packed away our stuff, and prepared the shuttle for our return, we gave the cabin a "cold soak." While we were confident that the thermal protection materials on the outside of the shuttle would do their job of protecting the structure (and us) from the 3,000 degree Fahrenheit temperatures that would ensue from the friction of reentering the atmosphere, we knew that all that heat would make it more difficult for us to maintain a comfortable interior temperature. So we cranked down the air conditioner to the point where it felt like we were inside a refrigerator. A side benefit of the frigid temperature was that a cool and comfortable astronaut is less likely to get airsick.

We strapped ourselves into our seats, and in parallel with our system checklist tasks, we started "fluid loading." We each drank a liter or more—depending on body weight and time spent in space—of salty water. Fluid loading would increase our blood volume and help counteract one of the most dangerous physiological changes that can impact astronauts on landing day: orthostatic intolerance. Astronauts experiencing orthostatic intolerance—the inability to remain standing upright—cannot maintain adequate arterial blood pressure and experience decreased blood flow to the brain when upright. This makes us more likely to become light-headed and perhaps even faint—not what you want to have happen while you're piloting a space shuttle careening toward Earth.

One hour later, while flying over a spot on the opposite side of the planet from the landing site at the Kennedy Space Center, we fired some tiny thrusters that flipped the shuttle around so that we were flying tail first, in position for the "deorbit burn."

Once in that tail-first position, the deorbit burn involves firing the shuttle's two most powerful thrusters, called orbital maneuvering engines, in order to initiate our deorbit—the process of pushing backward against the trajectory of our orbit to slow the shuttle down enough to take us out of orbit and start our descent into the atmosphere. Unfortunately, you can't just put your foot on a brake. The necessary maneuvering to slow down is precisely choreographed to get the shuttle to the correct angle and speed—slower, but not too slow—to safely enter the atmosphere. We only get one shot at the landing. If it goes too fast or too steep, the shuttle will burn up or break apart; going too shallow will cause the shuttle to "skip" off the atmosphere and miss the intended landing site altogether; going too slow lands us short of the runway and we crash.

A few minutes later, after the deorbit burn was completed, we flipped ourselves back around so that the shuttle's nose faced forward, to minimize the heat as we descended. About thirty minutes later, as we dropped from space, the density of the atmosphere started to increase; we were still at an altitude of about eighty miles (four hundred thousand feet) above Earth and about five thousand miles from the landing site; and thirty-one minutes away from landing, we were traveling at our top speed of Mach 25, or twenty-five times the speed of sound.

As we began our descent into the atmosphere everything seemed to happen very quickly. Traveling so fast through the atmosphere, the friction builds up on the outside of the shuttle, which becomes surrounded by a hot, glowing orange plasma that reaches up to 3,000 degrees Fahrenheit. Fortunately, the shuttle is covered with protective materials that keep the hot plasma from burning through the spaceship. It makes for a "fiery," yet surprisingly smooth, ride home.

As we continued to descend, the shuttle transitioned from flying like a spaceship to flying like a glider, and "pulling some Gs," we started to feel the force of gravity on our bodies again. (The term "Gs" refers to the force of gravity: one G is equivalent to the force of gravity felt on Earth.) At first, pulling just one-twentieth of a G, the change was barely noticeable, but about a minute later, as our commander (call sign "Scorch") called out "one-tenth of a G," I felt like I was being pushed into my seat and wondered how that could be just one-tenth. How impressive it was to be feeling this load of gravity on my body again. We rarely consider gravity in our daily lives here on Earth, but there was no denying its load on us as we returned from space. About ten minutes before landing, we reached our peak load of one and a half Gs, and we were back to one G by the time we touched down.

At an altitude of approximately 83,000 feet, having slowed from Mach 25 to Mach 2.5 (about 1,700 miles per hour), we were still about sixty miles from the landing runway and five and a half minutes to touchdown. As we continued to slice our way through the atmosphere, still traveling faster than the speed of sound, a double sonic boom (two distinct claps less than a second apart) could be heard across parts of Florida, signaling our imminent arrival.

The shuttle slowed to subsonic velocity when we were about twenty-five miles from the landing site. From this point and throughout the final approach, the shuttle descended toward the runway at more than ten thousand feet per minute (a rate approximately twenty times faster and about seven times steeper than a commercial airliner). Even though I knew we were traveling superfast and practically diving headfirst toward the ground, it wasn't until I caught a glimpse of clouds zipping by at the lower altitudes of our approach that I had any visual sensation of just how fast and at what a steep angle we were going.

Once the shuttle reaches subsonic speed and the commander takes over flying the ship manually, the shuttle circles and continues to steeply descend over the runway until it's lined up for the final approach. About seven miles out, with the runway visible from our front windows, we had slowed to about 400 miles per hour and were just a minute and a half away from touchdown.

We continued to dive toward the ground, but finally, at two thousand feet, the commander raised the nose sharply and slowed the rate of descent. At three hundred feet, the landing gear was put down, the shuttle crossed the end of the runway, and we touched down, having slowed to a landing speed of about 220 miles per hour.

The final part of the shuttle descent from forty thousand feet altitude to touchdown took just three and a half minutes. Most commercial airliners cruise at an altitude of about thirty-seven thousand feet, and when the captain makes the announcement that "we're beginning our initial descent for landing and we'll be on the ground shortly," usually that means you're still twenty minutes or more from actually landing. Now imagine if, by "shortly," an airline pilot meant only three and a half minutes!

After touchdown, I was astounded to realize that just one hour before I'd been in outer space. Now, just like that, I was back on Earth.

We landed the day after Thanksgiving. When the ground team opened the shuttle hatch and the outside air rushed in, I went into sensory overload—fresh, cool, clean air flooded the cabin, carrying the smell of dirt mixed with vehicle exhaust. The golden aura of the sunlight in late fall seemed different, yet normal. And I felt really, really heavy. So heavy that I laughed (and my crewmates Mike Foreman and Bobby Satcher laughed too) as I rolled out of my seat and onto the shuttle floor, so I could crawl in my big orange suit to the hatch.

Once there, I was greeted by the friendly and familiar faces of the recovery team, and with their encouragement, I mustered all the strength I had to lift myself from my hands and knees to standing. Yay! I say "yay" because the "simple" task of standing seemed impossible at the time. To do it on my own, I imagined I was in the gym and about to do the most difficult squat lift ever, and *voila!* I was up. I was thrilled to be able to walk myself off the shuttle and into our recovery vehicle. Still, my body felt so heavy that I even had to think about holding my head up. Our heads weigh a lot!

Once you're in the recovery vehicle, the post-flight testing begins. All the data that the medical and scientific research folks collected before and during the spaceflight is now compared to data collected post-flight, and they waste no time with that collection. Dr. Ed Powers and Cathy Dibiase, my flight doctor and nurse (and also my friends), were both on the recovery vehicle waiting to greet me. It went something like this: big hugs and smiles and "welcome home!" followed immediately by "Here's your urine collection bottle. The bathroom's over there. Do you need any help?"

This post-flight testing goes on for a few hours right after landing and continues with some regularity over the next few months.

(Actually, it goes on for the rest of our lives: I return to Houston each year for a thorough physical.) The doctors and scientists collect the data to study changes in our bodies due to our spaceflight, so that we can better understand how to care for astronauts during their time in space. The data is also valuable for figuring out what steps we can take before space travel to prevent the impacts of the space environment on our bodies, how we can improve countermeasures to those impacts during the mission, and how we can expedite recovery upon our return.

Every astronaut experiences their return to Earth a little differently. Some experience only mild effects on their balance and may feel some nausea, while others need to be carried from the spacecraft. There's no way to know how you'll respond in advance. Nothing about the way your body reacts here on Earth to the ups and downs of a roller-coaster ride or the roll and pitch of a boat at sea will tell you what your personal experience will be when returning from space. The one thing we all share, though, is feeling really heavy and tired. All I wanted to do was go to bed.

Fortunately, I'd received some good advice from my friend Peggy Whitson, a fellow astronaut and chief of the astronaut office at the time. Peggy was there on the recovery vehicle waiting to greet me with a big smile and hug, and she was the one who said to me, "Okay, Nicole, when you're done with all the testing they're gonna do on you and you get back to crew quarters with your family, I'm pretty sure you're just gonna want to sleep, but don't. Do yourself a favor and spend some time walking. Go to the gym and get on the treadmill. Set it to a low pace and just hold on and walk for twenty minutes, and I guarantee you, you will feel so much better when you step off."

To be honest, I wasn't excited about the idea of walking on the treadmill. When I got on it, I had to think intently about how I was walking and what was required to walk in a straight line. I was working hard just to hold my head up and stand up straight. Also, I discovered that it takes a while for those little hairs in our ears that help us sense which way is up or down to adjust to gravity too. Any motion that required me to move my head up and down made the whole world around me spin and left me with that feeling of needing to throw up but not being able to.

Still, I trusted Peggy's advice and did as she said, and dang it, she was right! I made my way with my husband and son as escorts to the small gym in crew quarters. We met my personal trainer Bruce (more on Bruce soon) at the treadmill, I walked for twenty minutes, and it worked! Almost miraculously it seemed, the little hairs in my ears recalibrated, and when I stepped off that treadmill, not only was I walking straight, but the nausea was gone too.

I was fortunate because in the broad spectrum of how differently people feel immediately after returning from space, I felt pretty good and was doing okay. Within the first couple of minutes, I could stand up and walk by myself; I could smile and have a nice conversation with the people who welcomed me home; I could get to the bathroom and take that first shower by myself; I was able to give my son and husband a big hug; and despite the initial nausea I felt that day, I was looking forward to the slice of pizza and my mom's famous green beans that I knew were waiting for me.

The day after landing at the Kennedy Space Center, the whole crew traveled with our families back home to Houston, and that same day I was in the gym to begin the forty-five days of intensive rehabilitation after my three months in space—something that every astronaut goes through after a long-duration spaceflight.

The astronaut corps is very fortunate to have a team of personal trainers. They're called astronaut strength, conditioning, and rehabilitation specialists (ASCRs), and we call them "acers." Bruce Neischwitz was my acer, and when I arrived at the first rehab session, he had set up a maze of little four-inch-tall orange cones on the gym floor.

"What's up with the Barbie doll cones?" I asked.

Bruce told me to stand next to him by the first cone. Then he said, "Here's what I need you to do. You're gonna hop over this first cone and..."

My laughter made it impossible for him to finish. "There's no way I can lift my body off the ground and hop over this cone!" I said. "The bottom halves of my legs feel like they weigh a hundred pounds!"

He just looked at me and smirked. "I'm telling you, Nicole, you can do it."

I smirked right back and tried to suppress my laughter, which made it come out more as a snort.

"Stop laughing," he continued. "And then you're gonna walk forward, hop over the next cone, walk sideways, hop over the next cone, walk backward..."

I continued to laugh and argue with him for another few minutes, but he was undeterred. Finally, I hopped over the first cone. To this day I'm not really sure how, except that Bruce *knew* I could do it, and so I found a way to believe it was possible.

If there's one thing that this experience taught me it's that even when we might not believe we can do something—like hopping over an orange cone after three months in microgravity—we always can do more than we think we can.

Thinking about the way our bodies can recover from the effects of flying in space, I realized the same is true for how quickly nature can recover if given the opportunity. During early 2020, at the beginning of the global COVID-19 pandemic shutdowns, we got a powerful sneak peek at how quickly nature heals itself. Scientists observed a sharp improvement in air quality, especially over quarantined regions. As an example, levels of one major air pollutant, nitrogen dioxide, over northern China, western Europe, and the United States decreased by as much as 60 percent in early 2020 compared to the same period in 2019. Nitrogen dioxide is a highly reactive gas produced during combustion, and it has many harmful effects on the lungs. The gas typically enters the atmosphere through emissions from vehicles, power plants, and industrial activities; as it reacts with other chemicals in the air it has far-reaching, damaging effects, like the formation of particulate matter, ozone (as discussed in chapter 2), and acid rain.

Such a significant and rapid drop in emissions was unprecedented. According to Jenny Stavrakou, an atmospheric scientist at the Royal Belgian Institute for Space Aeronomy in Brussels and coauthor of a May 2020 study for the American Geophysical Union, it had never happened even once since air quality monitoring from satellites began in the 1990s. Only the effects of short-term reductions in China's emissions—owing to strict regulations during events like the 2008 Beijing Olympics—have been comparable. These findings give us a glimpse of the potential future effects of more stringent emissions regulations on air quality. "Maybe this unintended experiment could be used to understand better the emission regulations," Stavrakou said, calling it "some positive news" in an otherwise "very tragic situation."[1]

While we should never downplay the tragedy of the COVID-19 pandemic, it has afforded us the opportunity to reconsider what our "new normal" will be going forward, as Dr. Stavrakou has pointed out. Instead of rushing to try to make up for what we "lost" in terms of productivity and profits, we have a chance to recognize and support the planet's ability to heal itself.

Our lives were uncomfortable and often difficult during the pandemic shutdowns, and yet, as we isolated ourselves in our homes and took other necessary precautions, the accompanying drop in manufacturing and transportation was responsible for cleaner air, increased wildlife activity in many areas, and perhaps even increased numbers of bees and butterflies in my own backyard. These shifts and others are examples of what's possible if we change our behavior, even a little bit. I'm hopeful that we can continue to explore alternatives to our old ways rather than just crank up the factories ten times over to try to make up some of the lost profits.

I'm encouraged when I read articles like this one from CNBC, "We Can't Run a Business in a Dead Planet: CEOs Plan to Prioritize Green Issues Post-Coronavirus."[2] This article points to three *modi operandi* from the pandemic that we can carry forward to make big gains in slowing and reversing climate change: more liberal work-from-home policies; reducing business travel and increasing the use of video calls; and making supply chains more local. The article is one of several indications that businesses are concluding that financial growth need not be stymied by a commitment to sustainability. All environmentally responsible shifts in our approach to business (as well as our individual approaches) will make life better.

The ISS motto, "Off the Earth, For the Earth," describes the greater mission of improving the lives of Earthlings by bringing the results of the work done in space back to Earth. The International Space Station program itself is arguably the most technically and politically challenging endeavor humans have embarked upon and executed successfully, and we built it in outer space.

While the ISS mission, with its engineering design and scientific focus, is primarily technical, a surprise benefit has been the outstanding model it has created of peaceful international cooperation. For humanity to continue to progress and prosper, we must design and build a future based on a peaceful and successful model of partnership like the ISS. We must base our earthly mission on the knowledge that we all share a singular vulnerability that requires us to protect our one shared home.

As crew members of Spaceship Earth, we must respond to the alarm bells our planet is ringing in order to protect our precious life support system. We already know how to measure the condition of our planetary life support system. In fact, we are already doing the measuring. Now we must use this information to not only understand our impact on the planet's ability to support life but to take action in response.

Many argue, and I'm one of them, that the solutions to many of Earth's greatest challenges can come from what we're learning and developing in space. Space research already has been crucial in averting one major environmental disaster: It was a NASA satellite sending back data that revealed the growing hole in the ozone layer over the South Pole. This information was acted upon and resulted in the Montreal Protocol, which was the first international agreement (and arguably the most successful one) to address a global environmental problem.

Space technology continues to provide the vital information needed to understand many of the most significant problems here on Earth. In September 2020, the World Economic Forum's Global Future Council on Space Technologies released a report, "Six Ways Space Technologies Benefit Life on Earth." The report opens with this statement: "The space sector's value to life on Earth is difficult to overstate."[3] The report addresses how space research supports or affects six areas—Sustainable Development Goals (SDGs), the climate, connectivity, global security, responsible business, and the global economy. In short, every aspect of life on Earth benefits from what's happening in space.

Much of the technology utilized in observing the Earth today was initially developed for probes sent to explore other planets in our solar system. Even the technology we're using to understand the history of the planets that we hope to one day populate as settlements offers us important strategies for climate change mitigation here on Earth. Space-based technology could help us move dangerous and damaging industry into Earth orbit and beyond, utilize the advantages of microgravity for manufacturing, and provide clean energy solutions like mining helium from the Moon or advanced solar cells and orbital solar power stations. Ideas written about as science fiction for decades have been and continue to be brought to life as scientific fact—brought back to Earth as solutions to some of our greatest planetary challenges. In the words of one of science fiction's foremost authors, Larry Niven, "The dinosaurs became extinct because they didn't have a space program. And if we become extinct because we don't have a space program, it'll serve us right!"[4]

I believe that space-based solar power is one of the most promising technologies. We already have all the know-how to shift away from fossil fuels and meet our energy demands by harnessing energy from the Sun. All our spacecraft and satellites, including the spacecraft that will take us back to the Moon and beyond, are powered by solar or nuclear energy. Solar power stations in space would continually face the Sun, beaming clean power back through targeted radiation to Earth, day and night, regardless of the weather. We have known for years how to move the generation of solar power off the Earth, but it is still seen as an "economically prohibitive" business because of the expense of getting the materials to space to build the power stations. (Or perhaps the problem is a "business as usual" mentality and a lack of will to make it happen. If in 2021 we can choose to enact a $1.9 trillion stimulus bill in response to the COVID-19 pandemic, why can't we choose to enact a roughly $20 billion space-based solar power development and implementation plan to enable a global energy solution?) There's no denying that this is a large-scale endeavor that will require complex integration, technology advances along the way, and perhaps even a couple of decades to fully implement, but think about it—we've done this kind of thing before. With the ISS!

Universities such as Caltech, with its Space Solar Project, already are developing scalable and modular technology for space solar power, and the Aerospace Corporation and other companies have called on the US government to work with industry and international partners to develop space solar power technologies.[5] In the United States, the Naval Research Laboratory (NRL) has taken the lead. In May 2020, NRL launched a space solar power experiment on the US Air Force's X-37B Orbital Test Vehicle-6. By 2025, China plans to demonstrate a one-hundred-kilowatt solar power station in low Earth orbit.[6] Ultimately, though, we, as a consortium of all humanity, will need to find ways to lift harmful industry off the planet. Let's hope we do it in time.

In an orbit about one hundred times farther out in space than the ISS are geostationary satellites that provide communications, weather monitoring, and continuous observation of our planet. A new ocean observation satellite, Sentinel-6 Michael Freilich, was launched on a SpaceX Falcon rocket from Vandenberg Air Force Base in California on November 21, 2020. The satellite was named in honor of Dr. Mike Freilich, the former director of the NASA Earth

Science division, who stated shortly before his death in 2020, "Earth Science shows perhaps more than any other discipline how important partnership is to the future of this planet."[7]

As an example of this type of partnership, the Sentinel-6 satellite is a joint NASA, European Space Agency (ESA), and National Oceanographic and Atmospheric Administration (NOAA) mission that will monitor ocean sea levels as well as temperature and humidity in the troposphere (the atmospheric layer in which we live). Measuring the height of the ocean gives scientists a real-time indication of the rate at which Earth's climate is changing. The oceans absorb about 90 percent of the excess heat from the planet's warming climate. Seawater expands as it heats up, resulting in about one-third of the average global sea level rise. Melting ice from land-based sources like glaciers and ice sheets accounts for the rest. The satellite will provide invaluable information for governments and local authorities tasked with planning for sea level rise and the resulting storms that will cause destruction of property, loss of life, and permanent displacement of affected populations.

As of this writing, twenty-eight Earth observation satellites are in orbit around our planet, measuring moisture, air quality, ozone, gravity, magnetic field, smoke, aerosols, methane, carbon dioxide, and nitrogen oxides, as well as providing high-resolution imagery of Earth. Sentinel-6 Michael Freilich is one of fifteen such satellites planned to launch through 2026. All enable us to measure the invisible threats to our planetary life support system and provide visible evidence of how our planet is changing over time.

In addition to the expected role that NASA and other government space agencies play in space exploration, the growing private-sector space industry also is dependent on the commercial enterprises and nonprofit organizations that are leveraging the benefits of space exploration, including the wealth of data provided by the Earth observation partnerships between NASA and international space agencies.

Motivated by profit as well as the desire to drive real solutions to our planetary problems, commercial space companies are developing new ways for us to get to space and opening up opportunities for more people to travel to space, and they are finding more profitable and beneficial ways to use the technology and information coming to us from space. For example, every time you use a navigation app, like Google Maps, you're using three different types of satellites—communications, remote sensing, and Global Positioning Systems (GPS)—to find your location and directions from point A to point B.

According to a 2020 report from the financial services company KPMG, the overall global economy of the space industry is roughly half a trillion dollars per year, and it's expected to grow to nearly a trillion by 2040. As of this writing, eighty countries have a space program, and the United States reports that at least 183,000 Americans are employed by the space industry.[8]

Companies like Planet Labs, whose motto is "using space to improve life on Earth," have deployed the largest constellation of Earth-imaging satellites in low Earth orbit in history. They are helping us increase our understanding of the world and improve life on Earth through higher-resolution data and imagery for industries like agriculture, energy, mapping, forestry, education, maritime, and more. They also partner with NASA to provide additional Earth observation data for measuring the World Meteorological Association's Essential Climate Variables (ECVs). An ECV is a physical, chemical, or biological variable, or a group of linked variables, that characterizes Earth's climate.[9]

Made In Space, Inc., the company that sent the first 3D printer to space for use on the ISS, is now leading the way in in-space manufacturing. Its innovations will not only help make space profitable but enable creation of a sustained human presence off Earth and utilize off-Earth manufacturing for the benefit of solutions on Earth too.

Nanoracks describes itself as "your portal to space" and has also adopted the motto "using space to improve life on Earth." Nanoracks was the first company to help commercial customers (companies, researchers, students) simplify the process for getting their research to the ISS. It envisions a near-term future when outposts are designed and constructed in space as factories, laboratories, greenhouses, and hotels that, by allowing more people to live and work in space, will not only improve life in space but on Earth as well.

Nonprofit organizations also are promoting the benefits of spaceflight and the use of space-based technology and information on a planetary scale. Two nonprofit organizations that stand out for me and are successfully using space-based resources in different ways to solve planetary problems are the Environmental Defense Fund and Geeks Without Frontiers.

For over fifty years, the Environmental Defense Fund (EDF) has been one of the world's leading environmental organizations, working to preserve the natural systems on which all life depends. It got its start back in 1967 as a small group of scientists and a lawyer from Long Island, New York, who were fighting to save osprey from the toxic pesticide DDT. Using scientific evidence, they got DDT banned nationwide. With more than two and a half million members and activists, the EDF continues to use science-based data to identify and implement solutions and policies in the areas of climate, energy, ecosystems, oceans, and health. Now the EDF is going into space. The organization is building a satellite dedicated to very precise, high-resolution measurement of global methane gas emissions. With a target launch date of early 2022, MethaneSAT is designed specifically to pinpoint the location and magnitude of methane emissions virtually anywhere on Earth. Cutting these emissions has been determined to be the fastest, cheapest thing we can do to slow the rate of global warming today, even as we continue to attack carbon dioxide emissions.

Geeks Without Frontiers is a younger and much smaller organization than EDF, but in twelve short years it has become an award-winning social enterprise with the goal of fostering a more connected world. It has established a platform for global impact by developing and promoting technologies for broadband connectivity that are helping boost standards of health, education, gender equality, and social and economic well-being for all of us, but most significantly for the estimated three and a half billion people around the world who remain unconnected to the internet. In 2019, Geeks was recognized by Space and Satellite Professionals International (SSPI) with the Better Satellite World Award for the group's innovation of combining satellite and smartphone technology to combat human trafficking, forced labor, and illegal fishing taking place off the coast of southeast Asia. This work was performed on behalf of the US Agency for International Development (USAID) and private-sector stakeholders.

Like the ISS motto "Off the Earth, For the Earth," the common theme of all these space agencies, commercial companies, and nonprofit organizations is a commitment to using space exploration and technology to find solutions to our greatest planetary challenges, so we can make life better. Every single one of these organizations is run by a team of brilliant people who believe they can make life better, and they believe that they can help solve the greatest challenges of our time. They all have big dreams that fuel their belief that we can build a better future, and I share their belief.

If you ever have the chance to see my friend Anousheh make a presentation, you'll hear her introduce herself this way: "I was born a long, long time ago in a country far, far away." While I'm not convinced that 1966 was such a long, long time ago, it is true that for many, her birthplace of Mashhad, Iran, could seem far, far away.

Anousheh Ansari and I first met in 2005 in Star City, Russia, where we both were training for our first spaceflight. She impressed me as a woman of true intelligence and a caring and infinitely positive spirit. She is petite and strong (I always feel ginormous next to her, but she could probably take me), and her naturally curly hair frames her beautiful face and eyes that always seem to be lit up in anticipation of her next adventure. I am thankful for her friendship and for the opportunity to share some of her story, as I believe she is one of the best examples of an Earthling who is doing all she can to "make life better." As usual, Anousheh was on the move during our interview, so we spoke by phone.

Anousheh was twelve when the 1979 Islamic revolution broke out in Iran, followed by an eight-year war between Iran and Iraq.

"It was a scary time," she recalled during our interview. "I started to hear gunshots, something I'd never heard before. There was a war going on around us, and many times my family and I had to take shelter to stay safe."

She still smiles when she speaks about finding refuge as a child through her love of the night sky. "In the summer, whenever I could,

I would sleep outside at night and look at the stars. I wanted to fly up and touch them, to feel what they were made of. During the war, my love of space and my imagination allowed me to travel to the stars, to someplace peaceful and safe, where I could take my family." Those nights under the stars ignited her dream to travel to the stars "for real" someday.

When Anousheh was sixteen, she emigrated with her family to the United States. Anticipating her arrival, she had imagined she would "attend StarFleet Academy and become a science officer on the *Starship Enterprise* and, as Mrs. Spock, would travel to space and explore all over the universe."

Well, to her dismay, and the equal disappointment of many others of us, there is no StarFleet Academy, but that didn't stop Anousheh. She didn't speak English when she arrived in the United States, but her dream of space exploration inspired her to become immersed in her education, and she earned degrees in electronic and computer engineering. She went on to become an entrepreneur and founded a successful computer technology firm, which she sold in 2001. The proceeds allowed her to support her love of space by invigorating the commercial spaceflight industry through the Ansari XPrize for Space, launched by the Ansari family in 2004, and by funding her own 2006 spaceflight, during which she became the first astronaut of Iranian descent, the first Muslim woman, and the fourth private explorer to travel to space and visit the ISS.

The Ansari XPrize was $10 million awarded to the first nongovernment organization to launch a reusable crewed spacecraft into space twice within two weeks. This challenge led to the formation of the space company Virgin Galactic and launched a new era for commercial spaceflight.

Anousheh says that the three chapters of her life should be titled "Imagine," "Be the Change," and "Inspire." She appreciates that imagination is a powerful and unique gift we have as human beings. She has always dreamed big and has never been afraid to take the actions necessary to pursue her dreams. She understands the importance of sharing her story as a tool for inspiring others to dream big.

Anousheh recalls that on the launchpad, strapped into the Russian Soyuz spacecraft with her crewmates, Russian cosmonaut Mikhail "Misha" Tyurin and US astronaut Mike "LA" Lopez-Alegria, "we put our hands together and said, 'Ready... here we go!' I thanked God for helping me realize my dream and for everything it has given me. I asked God to fill the heart of all beings with his love and to bring peace to this beautiful creation we call Earth."

Anousheh's Earthrise moment came when she caught sight of Earth through the window of her Soyuz spacecraft on the way to the ISS. "We were finally in orbit. We had been told that we can now open our belts, and it was when I was able to actually float up to the porthole next to my seat, and I looked out the window and the first thing I saw was Earth, and I was just awestruck."

We were on the phone together as we talked, so I couldn't see the look on Anousheh's face, but I could sense the emotion in her voice, the awe she was feeling again as she shared the experience with me.

"I was inside a capsule, and I couldn't feel heat or anything, but it was a feeling of this amazing warmth and energy, and I was smiling because I was very, very, very happy. But at the same time I started crying, it was so overwhelming. It was this mix of emotions. I'm like, 'Oh my gosh, if everyone could see this beauty, this energy.' This energy of life—it's so strong. I was like, 'If people could see this—life would be different on Earth.'"

Anousheh spent eleven days on the ISS. She made the most of every minute of her time there. She is known for posting the first blog dedicated to sharing a spaceflight experience. She posted in the days leading up to her spaceflight and then posted every day from space, sharing her experience with millions of followers on the ground. She also established her own routine on board the ISS. She told me, "My favorite thing was when everyone would go to sleep (because I had very limited time and I didn't know if I'd be back on Space Station), so when everyone would sleep and they'd turn off the lights, I would go and hover over the window that was next to me. I made sure that I had my sleeping bag next to a window. I would just stare out the window and watch Earth go by. I would just try to think about what's going on through the minds of people living their lives and just being consumed with the everyday noise of life."

She told me that she had an "undeniable sense of interconnectivity and interdependence. You connect with Earth as your home planet, so you become one with it, and that becomes home for you. I started noticing the beauty in nature and the changes of the weather and how it might be impacting things below me as I viewed it from space, and that has continued for me back on Earth. I felt like if we can elevate ourselves sometimes to see things from a different perspective, we can elevate ourselves beyond the noise and not get drowned by small things, and then we can really live our lives better."

We both agreed that the perspective from space is one we wish everyone could have, and we commented on how beneficial it would be if all our world leaders had the chance to experience this perspective—all floating together in front of a space station window gazing upon the awe and wonder of our unified planet below. If only!

Anousheh has been on a mission, even before her own flight to space, to make spaceflight accessible to more and more people because she knows that it's a transformational experience. All her work since flying in space has been built on her childhood dream of traveling to the stars. Remaining focused on this mission, she is also dedicated to inspiring young girls and women by sharing her own story.

Anousheh is involved in many different forums that support social entrepreneurship, and she is especially committed to ensuring the freedom of women around the world and supporting female entrepreneurs. She understands how her experience can be a model for people everywhere who are struggling to overcome economic and cultural barriers, and she is a proponent of STEM education and youth empowerment.

In 2018, she took on the lead role as the CEO of the XPrize Foundation, which has become "the world's leader in designing and managing incentive competitions to solve humanity's grand challenges."[10] Called "challenges," these competitions in the technology areas associated with space, oceans, learning, health, energy, environment, transportation, safety, and robotics offer the incentive of prize money awards. Funded by benefactors and sponsors, to date XPrize has awarded over $140 million toward its mission to enable social entrepreneurs to bring about radical global change through technology development for the sake of humankind and with the goal of bringing us closer to a better, safer, more sustainable world.

When Anousheh shared the news with me that she would be making the transition from CEO of her family company, Prodea Systems, to CEO of XPrize, it made perfect sense to me. To all who know the history, the success of XPrize was built on the model of the Ansari XPrize Space Challenge. It seemed a perfect choice for Anousheh to take on the leadership role.

Once Anousheh gets talking about the possibilities that she imagines from the work at XPrize, she can barely contain herself. I asked her about the influence of space exploration on life on Earth.

"Did you know how GPS was started because of [the Soviet] Sputnik [satellite]? I didn't," she said. "It was the curiosity of two scientists at the Johns Hopkins Applied Physics Lab who were looking at the beep sounds coming from Sputnik, and then their curiosity turned into a project that turned into GPS. GPS has transformed so many different parts of our lives, let alone that we can't go anywhere anymore without it. All of our communication systems, our banking systems, everything relies on the space program."

Anousheh also told me, "I'm so excited about the creative ways we can use space to get rid of big chunks of the carbon we're putting in our atmosphere." She described one way to reduce our carbon footprint: "I'm really interested in . . . putting big data centers in space. Take the emissions that result from refrigerating our data storage off the planet." To put this possibility in context, in 2016 it was reported that the world's data centers used more than Britain's total electricity consumption. Using 3 percent of the global electricity supply, and accounting for about 2 percent of total greenhouse gas emissions, data centers have the same carbon footprint as the aviation industry.[11]

"XPrize gives us a way to put these kinds of challenges out to the world and find really creative, audacious solutions."

Other XPrize challenges have included: $20 million for carbon dioxide transformation, $10 million for rainforest preservation, $10 million for avatar human transport, $7 million for ocean discovery (awarded in 2019 for deep-sea mapping), $5 million for AI solutions to global issues, $5 million for rapid COVID-19 testing, $5 million for reskilling the new workplace, and $1 million for literacy. Through her leadership role with XPrize, Anousheh is working every day to help improve life on Earth.

If we were given the opportunity to head back to space, I know that Anousheh would want to be strapped in right beside me, but we also know that you don't have to go to space to be able to make life better here on Earth.

To create a future where all life on Earth can thrive will require all of us working together to make life better. As Earthlings, we have the opportunity and the responsibility to step up and take on our role as the crew of Spaceship Earth, and to do so with the goal of improving life for all.

The significant hurdle that remains is that we, as Earthlings, as humanity, need to *choose* to do it. Just like the success of the ISS depends on all participants sharing a common mission, we have an opportunity here on Earth to work together and focus on our common interests. What greater shared interest could there be than our very survival?

We must choose to work together to improve life on Earth at the individual level as well as the collective level. I have highlighted examples in these chapters of both individual and large-scale efforts already in motion, many of which you can participate in, should you choose. Participating may be as involved as starting a movement or joining one, or as simple as picking up the next piece of trash you see on the ground and properly discarding or recycling it. Any of these actions, and many others like them, make life better for you and for everyone else too. The more of these actions we take, the more we become inspired to do. When we can see and feel the positive impact around us, we will choose to do more. We have the power to choose a path forward that ensures we not only survive but thrive.

Consider this: When we talk about "the fragility of our planet" and "saving the planet," what we really are concerned with is our own fragility and saving ourselves. While the planet is vulnerable in ways that can have serious implications for our survival, its 4.5-billion-year history has proven that Earth itself has survived much greater disruptions than the ones caused by humankind's presence over the past 200,000 years or so. Here's the simple math: we've been around for only about 0.004 percent of the lifetime of our planet. On the grander scale of our planet's total lifetime, we are a flea on the back of an elephant.

Now consider that it's only been since the beginning of the industrial age that our presence on the planet has led to the grand-scale degradation of our environment.

As the late comedian George Carlin famously said:

> The planet has been through a lot worse than us…hundreds of thousands of years of bombardment by comets and asteroids and meteors, worldwide floods, tidal waves, worldwide fires, erosion, cosmic rays, recurring ice ages.…The planet isn't going anywhere. WE are! The planet will be here for a long, long, LONG time after we're gone, and it will heal itself.[12]

While it's a harshly funny and in-your-face way of presenting our reality, Carlin's message is clear: We don't need to save the Earth. The Earth is much more resilient than we are. We need to save ourselves by protecting the things about our planet that sustain life. We need to change *our* ways—the way we treat the Earth and all the other life we share it with—in order to survive. The good news is that we have the technology and know-how to do what is necessary to counteract the negative impacts of the industrial age and amplify the benefits.

We need to shift the narrative from "saving the planet" to saving ourselves by doing what's necessary to heal and preserve the ecosystem that supports life, thus ensuring our survival. We must consider the significance of all life on the planet—from the tiniest life-forms to the atmosphere that wraps around us. Our future depends on our acceptance of the interconnectivity of our life with all other life on Earth and with the planet itself.

When students ask for my advice as they struggle to figure out what they want to do with their lives, I say, "Think about your life on Earth like an astronaut thinks about life on the ISS. Consider what it means to be a crew member, not a passenger, on Spaceship Earth. What will you do as an Earthling, on this shared planet and under the protection of the thin blue line, to help make life better for everyone? What will you do to ensure your crew's survival?"

I think the question we each need to ask ourselves is this: If you could have the future you want for yourself and your children and for all life on this planet, what would that future on Earth look like? I don't believe anyone would answer, "I want my children to live in a dystopian *Blade Runner* future." Nope—that's not the future we'd dream of for ourselves or future generations. Like Anousheh and her dreams of StarFleet Academy, many of us would answer the question with a sci-fi reference to a Star Trek future.

During our time in space, astronauts operate with what I like to think of as a fact-based optimism. In order to survive, we know that we have to rely on our crewmates, on the data we receive from our spacecraft, on the environment around us, and on our mission control team. Our reliance on the underlying science enables our optimism and drives the mission's success. We believe there is a solution to even the most challenging problems, and we approach these challenges with a "here's how we can," not a "here's why we can't," philosophy.

I will admit that after retiring from NASA, I found it a bit shocking to discover that the "outside world" doesn't necessarily operate with this philosophy in mind, even though all life on Spaceship Earth would fare best with this fact-based optimism. Whether you've traveled to space or not, we have all been given a view of our planet that should inspire an optimistic view of our future. The *Earthrise* photograph taken by Bill Anders has stood the test of time and is still the single best image to remind us all of who we are on this planet traveling through space together. This Earthrise perspective is the best one to consider when we are looking for the answer to the question: what does our future on Earth look like?

In these chapters, I've presented stories of people proactively addressing some of our greatest challenges. The following seven suggestions are meant, not as any kind of to-do list, but to point you toward ways of being that we have practiced successfully in the space program for decades. Think of them as a great boldface checklist for all Earthlings.

To address the planetary threats we now face:

- Act like everything is local (because it is)
- Respect the thin blue line
- Live like crew, not like a passenger
- Never underestimate the importance of bugs
- Go slow to go fast

- Stay grounded
- Whatever you do, make life better

Together we have the power to realize a future when life on our planet is as beautiful and peaceful as it looks from space. It's my wish that having read this book, you will take to heart (and mind) the reality that we all exist together on our planet and let it guide your actions each day to make life better.

We have the best planet. Please join me on her crew.

A view of my home state of Florida and the surrounding tropical waters as seen from the International Space Station. You can see some of the ISS solar panels in the picture frame.

NASA

CONCLUSION

IT WAS A NIGHT of new life and new hope.

On August 9, 2020, just after the full moon rose, scientists in scuba gear waited off the coast of Cook Island in the Florida Keys. With cameras and sample collection equipment in hand, the crew floated just above the ocean floor, surrounded by outcroppings of coral that had been planted as part of a restoration effort just five years earlier, and anxiously awaited a momentous release. Then, for the first time ever, they witnessed the spawning of restored massive star corals. The videos of this historic event show the spawning: a delicate blizzard of tiny white spheres, billions of coral sperm, and millions of coral eggs burst forth, ebbing and flowing in a beautiful ballet of snowlike flurries gracefully rising up the water column.[1]

The scientists were ecstatic.

Not only is coral spawning one of nature's most spectacular and mysterious events, but the synchronized phenomenon of coral sex signals new life and new hope for the viability of our coral reefs, and similarly, for humanity.

"Broadcast spawning" releases the zygotes and provides the greatest likelihood for the sperm and eggs to come together to form microscopic swimming larvae as the zygotes begin to rise up in the water toward the light at the water's surface. These tiny baby coral, known as planulae, "swim" in the water current for the next couple of weeks before they settle on the seafloor. They find something to attach to and, in a month or two, may grow to the just visible size of a pinhead. The new colony continues to grow very slowly at a rate of less than one inch per year.

This particular spawning of the massive star coral in Florida was especially significant because it was the first recorded spawning of corals that had been grown in a lab and replanted through a method called "micro-fragmentation and fusion," which had come about as the result of what my friend Dr. David Vaughan calls his "eureka mistake": the accidental discovery of a method to expedite coral growth and reef restoration.

As I began this book-writing journey, I decided to interview some of the people doing amazing work here on Earth to showcase what it looks like to live like a crew member on our planet. The very first interview I did was with Dr. Vaughan, an acclaimed marine biologist who has dedicated his life to turning that "eureka mistake" into a mission to protect our planet's life support system by restoring our coral reefs one polyp at a time.

I'd intended to feature this interview with Dr. Vaughan in chapter 1. I decided to save it for the end, though, when it became clear to me that his story not only illustrates the fact that everything is local, but touches on each of the themes in the seven chapters. In addition, there are interesting links between Dr. Vaughan's work and that of other people and organizations I've showcased.

In the spring of 2019, I arrived at the airport in Key West, Florida, excited to have the opportunity to meet David Vaughan in his natural habitat: the beautiful Summerland Key, just north of Key West. Dr. Vaughan greeted me with a big smile and a hug and asked me to call him Dave. He looked exactly like I'd expected from his photos—long, gray hair, a full beard, and smiling eyes. He was wearing flip-flops, a button-down shirt, and a wide-brimmed sun hat sporting the colorful logo of his nonprofit foundation Plant A Million Corals. He had the kind of laid-back air of comfort and confidence that you would expect in a lifelong surfer who feels right at home in nature. I could see that we would become fast friends.

Dave has been making pioneering breakthroughs and working at the cutting edge of aquaculture research and development ever since the early days of his career in marine science in the 1970s. As director of aquaculture at Harbor Branch Oceanographic Institution in Vero Beach, Florida, he created groundbreaking technology that allowed the sustainable scale-up of clam farming, which as of 2018 was a $40 million per year industry in Florida.[2]

Driving from the airport, Dave told me that clam farming had "boomed and made Florida the number-two clam-producing state in the US. I'm known as the guy who ruined the price of clams in New York and New Jersey."

He used some of the same technology from the clam farming system to develop a sustainable and environmentally friendly way to harvest ornamental fish (the kind we see in aquariums). This innovation provided an alternative to toxic techniques such as stunning fish in the wild with chemicals like cyanide; these practices not only are inhumane but also kill a large proportion of surrounding fish and damage coral reefs.

As a follow-up to the success of ornamental fish harvesting, Dave started to raise corals for the aquarium trade, and while doing so he had the chance to give a tour at Harbor Branch to Philippe Cousteau Jr. and Alexandra Cousteau, grandchildren of the undersea pioneer Jacques Cousteau. As Philippe learned about Dave's success with the coral, he stopped at one point, looked at Dave, and said, "You have it all wrong. Dave, if you can grow coral for aquariums, why not do it for the ocean reefs?"

Dave chuckled as he recalled this comment. "I felt like I was having one of those 'I could've had a V8!' moments," and he slapped his hand over his forehead to illustrate his reaction.

Shortly after this meeting, Dave explained, he left his position at Harbor Branch to join the Cousteaus at EarthEcho International and work with them on coral restoration initiatives.

From the Key West airport, we drove straight to the waterfront facilities of the Mote Marine Tropical Research Laboratory, where Dave began working in 2002 and had served as the founder and director of coral reef research from 2005 until he retired in early 2019. While at Mote, Dave focused on restoring coral reefs, which have been disappearing at an unprecedented rate as a result of climate change.

"Coral live beneath the surface, so for most people they are out of sight and out of mind," Dave said. "Our oceans provide us with irrefutable, visible evidence of the impact we're having on our planet's ability to support life."

One clear symptom is the devastating changes we're seeing with coral. In Florida and the Caribbean, coral cover has declined by 50 to 80 percent since the 1990s; in Australia, half the Great Barrier Reef was lost in the four years between 2014 and 2018.

I think we've all heard of coral bleaching, but like me, many might not really understand what it means. Turns out that coral is extremely sensitive to changes in temperature. As ocean temperatures rise, the algae inside coral starts to produce oxygen faster, and because the coral animals can't get rid of the excess oxygen fast enough, they eject the algae. The problem: algae is their food source.

"Coral would rather starve to death than overdose on oxygen. After expelling the algae, it turns white and then dies, which is known as 'coral bleaching,'" Dave explained.

I am a lover of the undersea world. I grew up on Florida beaches, I'm an avid scuba diver, and I spent eighteen days on the Aquarius undersea habitat in preparation for spaceflight. Besides enjoying opportunities to explore the undersea world, I have long been an advocate for its protection—maybe my love for the undersea world is *why* I became an advocate. But my time with Dave showed me I had much to learn.

As we toured the Mote facilities, my coral reef lesson began with Dave showing me pieces of coral growing in the laboratory tanks. "Coral is an animal, plant, microbe, and mineral that all live together in a symbiotic relationship," he told me.

At the time, I kept quiet about my ignorance about these facts but later fessed up that, while I knew there was an animal involved somehow, like most people I thought coral was a big rocklike structure. Turns out that the animal is a coral polyp that's made of tiny clear tissue with even tinier tentacles and a mouth. The plant (algae) lives inside the polyp and photosynthesizes to produce oxygen and also provides the polyp with nourishment and color. The microbes (bacteria) surround both the polyps and algae and provide an antibiotic-like, protective immunity for the system. All grow together as a cloned colony that secretes calcium carbonate to build a shared rocky skeleton.

I was shocked by how little understanding I had of the complexity of coral, and of just how crucial a role coral plays in the survival of life in our oceans—and hence in our survival too. Oceans cover two-thirds of the planet, and coral reefs cover less than 1 percent of the ocean's floor, yet they support up to 50 percent of the world's fisheries.

"Coral reefs are like an oasis in the desert for marine life. They provide a barrier that protects our coastlines, and they provide significant economic benefit through fishing, diving, snorkeling, and tourism," Dave said. Florida's reefs alone are estimated to have an economic value of $8.5 billion annually and generate over seventy thousand jobs.[3]

Dave then looked me square in the eye and asked, "Why should we care about coral?"

I rambled with a couple of semi-intelligent reasons based on what he had already told me, to which he very patiently and politely replied, "Do you like to breathe?"

"Of course!" I said. "Who doesn't?"

"Well, coral reefs are like underwater rainforests," he said. "Fifty to eighty percent of the oxygen we breathe comes from the ocean through marine plant sources like phytoplankton and seagrass, and through the algae in coral. At least every other breath you take is from the ocean and its coral."

While working in the very same lab that we were visiting, Dave described to me how he and his Mote colleague Christopher Page, a biologist, serendipitously discovered a way to grow coral up to forty times faster than in the wild. Not only that, but this coral happens to be better able to tolerate the higher temperature and acidity that exists in our oceans today. It all began when he pioneered a lab technique for encouraging corals to spawn.

In the Mote lab, they made what Dave called "the first test-tube coral babies." The process was revolutionary and successful, but also "disappointingly slow." Dave explained: "It took one year to grow coral from the microscopic baby larvae to just the size of a pencil eraser. It was another three years before it reached the size of a quarter or even started to look like coral."

"We realized that this would not be an effective way to restore coral reefs," Dave said. The Mote team gave up on the technique.

As they ramped down these experiments, Dave was cleaning up the lab and had to move some of the coral pieces from one location in the tank to another. "Several small pieces broke off, and tiny polyps fell to the bottom of the tank," Dave said. "I thought, *Well, they're toast.*"

A few weeks later, when he passed by the tank, Dave noticed that the individual polyps had multiplied twelvefold and grown to the size of the original piece they'd broken off of.

"I knew I had to figure out what happened," he said with excitement in his voice. "We repeated the 'mistake' by breaking off tiny pieces of coral and watched them grow in just weeks to a size that would normally have taken years. It turns out, coral is slow-growing unless it's damaged. A wonderful, natural healing process. Eureka!"

Because Dave and his team are growing species that are more tolerant of changing conditions and rising ocean temperatures, it is now possible to replant these small pieces of coral together on a dead coral head in the ocean and, in just a couple of years, restore a VW Bug–sized coral that would have taken 250 years to grow naturally.

Like many other accidental scientific discoveries that have changed the world for the better, such as X-rays, the pacemaker, penicillin, and insulin, Dave's "eureka mistake," now known as "microfragmentation," is changing the world of coral reef restoration.

For a planet that's had more than 75 percent of its global coral reefs bleached since 2014, Dave's work provides not just hope but a simple and effective solution for restoring our reefs faster than nature can on its own.[4]

"We can replant the massive, boulder-sized brain and mountain corals that build the structure of the reef. They're like the giant maples and oaks of the forest. This is game-changing, for the coral reefs, and for us," Dave said.

In early 2019, when Dave retired from Mote, he and his wife Donna founded Plant A Million Corals, a nonprofit foundation with the mission to, well, plant a million corals.

Before Dave's "eureka mistake," it would have taken six years to produce six hundred coral for planting, but now it's possible to grow six hundred coral *each day*. Between his work with Mote and partnerships with organizations like NOAA and the Nature Conservancy, by 2019 over two hundred thousand corals had already been planted in the Florida Keys and across the Caribbean. By sharing the process and scaling to implement more coral restorations in the Keys and around the world, planting a million corals could be realized in two years with a price tag as small as $1 million.

After two years and a million corals planted at one dollar each, our oceans' reefs could be on the way to recovery. Seems to me this is a no-brainer, and something we should support right now in response to the emergency alarms that the world's corals have been sounding for decades.

Through the work of Plant A Million Corals, Dave and his team are able to carry forward his laboratory discovery through independent coral restoration activities in places beyond the Florida Keys, including the Virgin Islands, Bonaire, Belize, and Costa Rica. By working in ongoing partnership with other organizations like the Nature Conservancy and Fragments of Hope, the foundation will be able to extend its coral restoration activities even further around the planet.

Dave's story reminds me of the connections between all the other people I interviewed for this book and the inspiration I drew from them. All are on a mission to use their life experiences, their talents, and their time to improve life on Earth. While none of them have met Dave Vaughan, their missions and stories are connected to his, and they know about one another now. Just as the work of all these people connects to Dave Vaughan and his coral restoration efforts, his story also exemplifies each of the lessons—the ways of being—described in these chapters.

The degradation of coral reefs is impacting not just the coastal communities where the reefs are located but all of us. Earth is one planet, so **everything is local**. All life depends on the health of these ecosystems.

The coral reefs, in partnership with the overall ocean environment, are critical to the cycles of our atmosphere and the absorption of greenhouse gases—the key to maintaining the **thin blue line** that wraps our planet and protects us all from the deadly vacuum of space.

The coral species that make up the massive coral reefs are among the trillion estimated different species of life on Earth. Regardless of how small these tiny polyps are, **they have a role to play as crew**, and we should **never underestimate their importance** in sustaining the diversity of planetary life.

Dave's discovery of the micro-fragmentation method and his work on coral restoration provide wonderful examples of the **"go slow to go fast"** approach. Once we know what the problem is and we have devised a great solution to solve it, we need to act swiftly, while there is still time, to bring things back into balance.

Like the coral, we as individuals, as humanity, as a planet, in one way or another are all in need of restoration. The life that surrounds and depends on a healthy coral reef also presents a stunning display that sparks awe and wonder. To be in the midst of it is **grounding** and can bring a greater understanding of the interconnectivity of all life on our planet.

And everything about the work that Dave and everyone I interviewed is doing is absolutely driven by the motivation to **make life better**.

These unexpected connections demonstrate the power within each one of us to take actions toward overcoming our greatest challenges, like climate change, that contribute to millions of other actions. Everything counts.

In concluding, I've focused on Dave's coral restoration story because I believe it mirrors the overall motivation behind this book—to remind us all of the awe and wonder that surrounds us every day; to confirm the three simple truths that we all live on a planet, as Earthlings, protected by the same thin blue line; and to inspire each of us to take consistent action to do whatever we feel called to do to make life better.

At the end of our day together, with the interview completed, Dave and I sat on the deck of his house with his family for dinner and relaxed. As the sun was setting in its most spectacular Florida way, Dave pointed across the canal to where the Mote Marine Lab is located and where he had worked all those years, and he casually mentioned to me that he had paddled the distance across the canal to work every day while holding his breath.

"Just another way for me to reduce my own carbon footprint," he said. "The little things add up."

As we sat together in appreciation of the beauty that surrounded us, Dave continued: "Our planet is misnamed. It really should be called 'Planet Ocean.'"

He squinted at the last glint of sunlight as the Sun sank to the edge of the horizon.

"You know, there are so many analogies between outer space and inner space. You've had the luxury of going out into space and looking back, and I've had the luxury of going underwater to a few thousand feet and coming up and seeing the same thing from down under. What a big ocean. What a big universe."

All astronauts I know have shared the impact of the Earthrise perspective on their own lives: how their minds and hearts were opened by a view of our home through a spaceship window. I wish that everyone could have the experience of seeing Earth from space, as we all could benefit. I'm thankful for Dave's acknowledgment of the similarities between his inner space and my outer space perspectives. He so clearly understands one of the most important messages I want to share—that you don't have to see Earth from space to experience an Earthrise moment.

Only a few of the people I interviewed had ever seen Earth from space, yet all of them had had an Earthrise moment, even if they didn't call it that. An Earthrise moment wakes you up to the reality and significance of who we are and the fact that we all are together on our one planetary home in space, and that we all share the responsibility that goes along with it. In fact, each of my astronaut friends shared with me that they had experienced multiple moments like this since returning to Earth because they are open to the awe and wonder that surrounds them.

The people I interviewed for this book had so many profound insights about our connections to the Earth and one another as they graciously shared their Earthrise moments with me. As they did so, I experienced what felt like a spark of inspiration, an Earthrise moment, with each of them.

Serena Auñón-Chancellor commented that, while performing a cellular biology experiment on the ISS, she'd always remind herself as she cared for the cells that they were "the tiniest form of life... like little crew members living with us." She was amazed that the warmth of their cell chambers "felt like home" to her, giving her a connection to Earth as her home.

Rowan Henthorn, while exploring the farthest reaches of the Pacific Ocean, would look up at the stars at night and consider "what this tiny boat of girls traveling together across the ocean must look like from far out in space"; at the same time, her "whole world also felt like just the boat and the five meters of water that surrounded them at any time." This gave her a new understanding that "there is no *away*," and that we all really are connected.

Guy Laliberté pursued his time on the ISS as a "humanitarian space explorer" with a mission to raise awareness of the water issues affecting so many people on Earth. He saw a planet that "stands like a beacon in the darkness," but whose life support system is also threatened by its inhabitants. He simply asks his fellow Earthlings, "How about we help her out?"

Scott Harrison reminded me that an Earthrise moment isn't always beautiful—sometimes it's the shock of being suddenly aware of the vastness of human suffering. He said that on his first trip to Africa he realized that he'd "never been confronted with any kind of mass or swarm of human suffering in aggregate like that before." Scott was stunned that so many people lacked access to clean drinking water, which he had so casually charged people "$10 per bottle for" at the events he used to host in New York City. Like Guy, Scott's Earthrise moment shook him into action, and connecting people with the Earth's lifesaving resources has become his life's work.

David Liittschwager experiences Earthrise moments all the time, through his "desire to understand the way the world is made." He would love to do a story about "what exactly it took the world to make a flower." He always looks forward to that "tingly, lovely feeling on the back of my head when I see or hear something beautiful," like making eye contact with an octopus and appreciating the awe and wonder of "mutual regard."

Mark Tercek spoke of "discovering my inner environmentalist" while traveling in Costa Rica with his family. "There's an inner environmentalist in everyone waiting to be discovered," he said. To overcome a planetary challenge like climate change, he believes that "you need the whole world to play ball…to find our common ground." "Hey, people," he suggested, "like astronauts from space, you can look at the planet too in different ways."

Dan Good's overall perspective on life and how to live it with purpose came into focus when he woke up to his preacher's admonition: "We all should be living life with the knowledge that it's a gift and privilege to be alive."

Looking out her spaceship window, Anousheh Ansari was awed by the view, which made her feel like she was "witnessing the energy of life." She thought to herself, *Oh my God, if everyone could see this beauty, this energy... life would be different on Earth.*

I recently saw pictures from a climate march, and one marcher held a sign that read, THE CLIMATE IS CHANGING, SO SHOULD WE. Like Dave Vaughan, each of the people I interviewed is doing something with their Earthrise moment—making changes in their lives and taking action in response to it. They are behaving like crewmates.

In a final reflection, Dave said, "If everybody could just get a few minutes each day of connecting themselves to nature and the whole cosmos... whether that's seeing a sunset, watching a star come out every night, or going into the woods to see the cycle of creeks and rivers and plants, I think we would do a lot more to preserve what we have."

I couldn't have said it better myself.

We all can take Dave's lead, not only appreciating the nature that surrounds us, but being aware of our impact on our environment and taking personal action to make things better.

When I think of a coral reef and how it grows, I feel a great deal of hope for humanity. Each coral polyp on its own is a tiny dollop of a life-form, yet each one is critical to developing a mighty reef. This is a spectacular example of how a collective of seemingly insignificant individuals can gather forces and bond together and create something bigger than themselves. I believe that we humans can do the same.

The monumental coral spawning that took place on that moonlit night in August 2020 in the Florida Keys happened just a year after my interview with Dave and only a few miles off the coast from where we spoke. I reflect on this with awe and wonder. This coral had been grown at unprecedented speed using a method that came about as a result of Dave's "eureka mistake." The coral was planted in the ocean by humans and had spawned on its own in a fraction of the time it would have taken to reach maturity without the benefit of microfragmentation. (This is significant because for coral, size does matter. Sexual maturity for coral, the ability to spawn on their own, is a factor of size not age.) All this was made possible because a few humans chose to behave like crewmates here on Spaceship Earth.

As Earthlings, we each can choose, must choose, to do our part to overcome our greatest challenges. We can view the work on the ISS as a model of what it looks like to work together, regardless of race or gender or country of origin, as one crew on one mission: to overcome the challenges that threaten the survival of all life on Earth. We can take the ISS motto, "Off the Earth, For the Earth," and bring it back to Earth. We can see ourselves for who we truly are: Earthlings. Earthlings who share a planet spinning in space, who have an opportunity and the ability to bring new life and new hope to our shared future.

ACKNOWLEDGMENTS

"SPACE. THE FINAL FRONTIER."

Captain Kirk shares these iconic words at the beginning of every *Star Trek* episode. They are part of my earliest childhood memories and have stayed with me my entire life. I think they have remained important to me because the show, and these words, inspire a vision of a positive future for life on (and off) Earth. I'm thankful for the image of our future put forth in *Star Trek* because it helped me appreciate the power of transforming imagination into reality. I think it helped us all believe that it is possible for humans to overcome seemingly impossible challenges—to even travel to space on missions like Apollo to the Moon, on the Space Shuttles, and to live on the International Space Station.

I want to thank the Apollo 8 astronauts for sharing their extraordinary view of Earth from space with all of us. Their *Earthrise* image is still the most compelling presentation of our one shared home in space.

For me, the measure of the significance of any journey comes down to the people we share it with, and the best way to end an amazing journey is always to come home. Traveling to space was an amazing journey, and I'm thankful to bring my experience back to Earth. My heartfelt thanks go out to all (and there are so many) who lifted me up along the way, who shared the journey with me, and who welcomed me home.

If I mentioned you in the body of the book, please accept that mention as my acknowledgment and as a sincere expression of gratitude for your support, your friendship, and the enthusiasm with which you have taken on your role as a fellow crew member here on Spaceship Earth.

For their help along my path to becoming an astronaut, I am hugely grateful to Jay Honeycutt, Tip Talone, and Ken Cockrell, who answered my phone call back in 1997 and were willing to be references on my astronaut job application. (They were even willing to do it again two years later.) I hope they know that I wouldn't have even been in a position to ask for their references if not for what I learned from each of them while working as a NASA engineer at the Kennedy Space Center.

I'd like to thank George Abbey for his friendship, and for everything he's done and continues to do for our human spaceflight program. In the 2018 biography of George Abbey by Michael Cassutt titled *The Astronaut Maker*, I discovered so much more about George. I highly recommend this book for anyone interested in the history of NASA and the selection of astronauts. One of the most surprising things I discovered from this book, and something so central to the *Back to Earth* story, is that George was the NASA employee responsible for selecting the image from those taken by the Apollo 8 crew that became known as *Earthrise*. I remember reading this the first time and just staring at the page. And then reading it over and over a couple times before continuing. I know if I ever were to say to George, "You never told me you selected the Earthrise image," that he would just look at me and smile and say in his quiet, rumbly voice, "Well, you never asked."

My Bug astronaut class of 2000 knows there wouldn't have even been an astronaut class of 2000 without George Abbey. Thank you, George.

To all my Bug classmates, and spaceflight and undersea crewmates, and their families, as well as all my fellow astronauts, cosmonauts, and crew support and training and mission control friends around the world—thank you for making "being an astronaut" about something so much more important than just flying in space.

As I've transitioned from being an astronaut to my post-NASA life, I'm thankful to my Space for Art Foundation team—Ian, Loli, David G., Gordo, Alena, and David D.—for their friendship and for helping me discover my next mission in life and become a better Earthling. We have so much more good work to do together. To our Constellation crew—Guy, Christoph, Jeremy, Jacob, Jan, Anousheh, Leland, and Ron—I look forward to working with all of you as we continue to spread the power of the Earthrise message around the world. To Frank White for enlightening all of us with your Overview Effect philosophy. To Amanda Lee Falkenberg, I'm so thankful for the serendipity of our friendship and for the opportunity to help bring your extraordinary "MOONS SYMPHONY" to life as we push ahead on the *Back to Earth* mission.

A special thanks goes out to my two heroes of spaceflight and art who came into my life as mentors and friends as I was preparing to retire from NASA—astronauts and artists Alan Bean and Al Worden. We lost both of these beautiful Earthlings way too soon (2018 and 2020, respectively). Alan Bean was the fourth person to walk on the Moon during Apollo 12; after retiring from NASA, he became a full-time artist committed to sharing his spaceflight experience with the world through his paintings. Al Worden was an Apollo 15 crew member, and as command module pilot, he circled the Moon alone for three days. Al also performed the first deep-space spacewalk and was a poet. Not only was my transition from astronaut to artist made easier through the friendship and support of these two men, but life on Earth is better because of them.

This book would not exist if not for the amazing people I call my "book people." At the top of the list is my husband, Chris, who convinced me that this was an important book to write and who remained my number-one cheerleader, somehow graciously putting up with my "behavior" as I often struggled to get on a roll with the actual writing. Thank you to my friend Will Li, who introduced me to his friend and book coach Robin Colucci. (Do yourself a favor and read Will's book, *Eat to Beat Disease*.) I call Robin my "one-woman mission control team." Thank you, Robin, for being there every step of the way with your guidance and belief in this story, for working your magic and helping me bring the story to life, and for the friendship that's grown out of it all. Thanks to Will Li again for connecting me to the literary legal genius John Taylor "Ike" Williams. Thank you, Ike, for guiding me through the contractual side of the book-writing process, which gave me the comfort to concentrate on the writing itself. I enjoyed our conversations and hope you enjoy the book.

Thanks to my friends Anousheh Ansari, Simone Giertz, Robert Kurson, Jane Root, Lynn Sherr, Frank White, and John Zarella, who provided me with endorsements when this book was nothing more than a paragraph describing the idea. It means a lot to me that you believed in my ability to tell this story. Thank you, Robert, for inviting me, as a rocket woman, to write a short blurb for your wonderful book *Rocket Men: The Daring Odyssey of Apollo 8 and the Astronauts Who Made Man's First Journey to the Moon* (another book I highly recommend), for offering to help me when the time came to write this book, and for introducing me to the team at Sterling Lord Literistic, who welcomed me as a client and assigned Jenny Stephens as my literary agent.

Jenny, I am so grateful for your patience in helping me navigate the mysterious and previously unknown-to-me world of book publishing. From our first phone call, I've felt a thoughtful and comfortable connection with you. Your sense of who I am and your understanding of my goals with this book were instrumental to finding a home for *Back to Earth* with the outstanding publishing team at Seal Press and the Hachette Book Group.

Another special thanks goes to Emi Ikkanda, my editor at Seal Press. Thank you, Emi, for your excitement, guidance, and questions, which helped me draw out an even more compelling story. From our first meeting at the Hachette offices in New York City, it was clear that you had thoughtfully read my proposal, understood the significance of my story, and were genuinely excited and knowledgeable about space exploration and science. I'm also thankful to Emi for inviting the brilliant Roger Labrie to our book team and to Roger for sharing his masterful line editing expertise with me. Thank you, Roger, for reviewing the manuscript with fresh eyes! And to Melissa Veronesi and Cindy Buck for the meticulous attention to detail during the copyediting phase and helping me to complete such a high-quality manuscript to send to print. I'm also grateful to Liz Wetzel, Jessica Breen, Kara Ojebuoboh, and the whole publicity and marketing team for developing a beautiful strategy for the promotion of the book. I am in awe of the excellence that all my "book people" bring to the *Back to Earth* team.

As I mention in the book itself, I am not a scientist, but I played one on a space station. In support of the research for the book I'm indebted to my scientist friends and those of you I sought out or met and now call friends—thank you, Liz Warren, Roger Weiss, Will Stefanov, Sue Runco, Erin Anthony, Rachel Barry, Chris Hickey, Michael Rodriggs, Dorit Donoviel, Rachael Dempsey, Gary Strangman, and Dennis Paulson. You all responded thoughtfully to my out-of-the-blue and sometimes frantic emails with questions about the science behind the story.

A special thank-you is in order to the UNESCO Biosphere Isle of Man team. I already loved the Isle of Man, but Jo Overty, Richard Selman, Sophie Costain, Sarah Mercer, Nikole Cervantes, and Rowan Henthorn all graciously shared even more with me about this unique and wonderful place. I hope you are happy with how I shared the story of the island's well-earned honor as the only country designated as a UNESCO Biosphere.

Thanks to my friends Tim and Susie Barth, Ron and Carmel Garan, Michael and Margaret Potter, Jamie Jarvis, Mary Baysore, Julie Finger, Alyson Hickey, Terry Lee, Auntie Carol and Uncle Peter, Jay and Peggy Honeycutt, Tip Talone, Mike and Jane McCulley, Joe and Maureen Rhemann, Jenny Lyons, David Martin, Kevin Mellett, and Nadya Kalinina; thanks also to Greg Rocktoff and Liz, Bryan, James, and Anne Stott, the best in-laws a girl could ask for. You all are the people I feel lifting me up whether you are near or far.

And last but never least, a big thank-you to my family. To my parents, who shared what they loved with me and set me on this path. To my dad, who gave me those first views of life from an off-Earth perspective, and to my mom, who has supported me every step of the way with strength and love. To my sisters, Shelly and Noelle, for the meaningful paths they've chosen in life, for their love, and for the love they demonstrate every day in how they care for their families. (And thank you, Noelle, for traveling with me to Key West and for transcribing all the interviews.) To my husband Chris and my son Roman, thank you for *everything*.

And finally, thank you, dear reader, for choosing to read my book. I hope you are glad to have read it.

HOW TO GET INVOLVED

IN THIS BOOK, I've shared the stories of inspiring change-makers who have all taken on their important role as crewmates here on Spaceship Earth. Please visit www.backtoearthbook.com for additional information about these individuals and their organizations, as well as some others I've discovered and support. You'll also find some information and ideas that I've found useful as I continue to pursue ways to improve my own skills as a crewmate and Earthling. I hope you will find this helpful too, and that you'll use the website as a space to share your own ideas for our shared mission to protect our planetary home.

NOTES

CHAPTER 1: ACT LIKE EVERYTHING IS LOCAL (BECAUSE IT IS)

1. Bill Anders, "50 Years After 'Earthrise,' a Message from Its Photographer," Space.com, December 24, 2018, www.space.com/42848-earthrise-photo-apollo-8-legacy-bill-anders.html.

2. United Nations, Sustainable Development Goals, "UN Report: Nature's Dangerous Decline 'Unprecedented'; Species Extinction Rates 'Accelerating,'" May 6, 2019, www.un.org/sustainabledevelopment/blog/2019/05/nature-decline-unprecedented-report/.

3. Josie Glausiusz and Volker Steger, "The Intimate Bond Between Humans and Insects," *Discover*, June 25, 2004, www.discovermagazine.com/the-sciences/the-intimate-bond-between-humans-and-insects.

4. EarthSky, "How Much Do Oceans Add to World's Oxygen?" June 8, 2015, https://earthsky.org/earth/how-much-do-oceans-add-to-worlds-oxygen.

5. Florida International University Institute of Environment, "Facilities and Vessels," https://aquarius.fiu.edu/dive-and-train/facilities-and-assets/aquarius-undersea-laboratory/index.html; NASA, "About Aquarius," March 28, 2006, www.nasa.gov/mission_pages/NEEMO/facilities.html.

6. Kaneda Toshiko and Carl Haub, "How Many People Have Ever Lived on Earth," PRB, January 23, 2020, www.prb.org/howmanypeoplehaveeverlivedonearth/.

7. Tony Phillips, "Space Station Astrophotography," NASA Science, March 24, 2003, https://science.nasa.gov/science-news/science-at-nasa/2003/24mar_noseprints.

8. James D. Polk, principal investigator, "Vision Impairment and Intracranial Pressure," NASA, www.nasa.gov/mission_pages/station/research/experiments/explorer/Investigation.html?#id=1008.

9. Karina Marshall-Goebel, Steven S. Laurie, Irina V. Alferova, et al., "Assessment of Jugular Venous Blood Flow Stasis and Thrombosis During Spaceflight," *JAMA Network* 2, no. 11 (November 13, 2019), https://jamanetwork.com/journals/jamanetworkopen/fullarticle/2755307; "LSU Researcher Was Lead Author of Study on Astronaut Blood Clot Risk," *The Advocate*, January 3, 2020, www.theadvocate.com/article_59b85012-2e5d-11ea-9c92-d3cd7a667555.html.

10. "How Does Spending Prolonged Time in Microgravity Affect the Bodies of Astronauts?" *Scientific American*, October 6, 2003, updated August 15, 2005, www.scientificamerican.com/article/how-does-spending-prolong/.

11. NASA, "Space Station Research Explorer," www.nasa.gov/mission_pages/station/research/experiments/explorer/; ISS Program Science Forum, *ISS Benefits for Humanity*, 3rd ed., NP-2018-06-013-JSC, www.nasa.gov/sites/default/files/atoms/files/benefits-for-humanity_third.pdf.

12. Serena Auñón-Chancellor, "Angiex Cancer Therapy in Space," NASA, YouTube, August 21, 2018, www.youtube.com/watch?v=AyfMCNfcWSc&feature=youtu.be.

13. Deborah Borfitz, "Space Is the New Frontier for Life Sciences Research," *BioIT World*, September 16, 2019, www.bio-itworld.com/2019/09/16/space-is-the-new-frontier-for-life-sciences-research.aspx.

14. "Crystallizing Proteins in Space to Help Parkinson's Patients on Earth," ISS National Laboratory, April 2, 2019, www.issnationallab.org/blog/crystallizing-proteins-in-space-to-help-parkinsons-patients-on-earth/; Allison Boiles, "LRRK2 Science in Space: 'The Key to Curing Parkinson's May Be Out of This World,'" Michael J. Fox Foundation, January 18, 2019, www.michaeljfox.org/news/lrrk2-science-space-key-curing-parkinsons-may-be-out-world.

15. Michael Johnson, "Keeping an Eye on Algae from Space," NASA, December 11, 2019, www.nasa.gov/mission_pages/station/research/news/b4h-3rd/eds-keeping-eye-on-algae.

16. Henry George, "The Unbounded Savannah," in *Progress and Poverty* (1879); Adlai Stevenson, "Strengthening the International Development Institutions," speech before the United Nations Economic and Social Council, Geneva, July 9, 1965.

17. Buckminster Fuller, *Operating Manual for Spaceship Earth* (Carbondale: Southern Illinois University Press, 1969), 54–55.

18. "Learn About Earth's Spheres," Generation Genius, www.generationgenius.com/earths-spheres-for-kids/.

CHAPTER 2: RESPECT THE THIN BLUE LINE

1. Michael Johnson, "Plant Growth on the International Space Station Has Global Impacts on Earth," NASA, March 10, 2020, www.nasa.gov/mission_pages/station/research/news/b4h-3rd/hh-plant-growth-in-iss-global-impacts.

2. Mark Garcia, "Space Debris and Human Spacecraft," NASA, September 26, 2013, www.nasa.gov/mission_pages/station/news/orbital_debris.html; Maya Wei-Haas, "Space Junk Is a Huge Problem—and It's Only

Getting Bigger," *National Geographic*, April 25, 2019, www.nationalgeo graphic.com/science/space/reference/space-junk/.

3. Elizabeth Bourguet, "Despite International Efforts, Scientists See Increase in HFC-23, a Potent Greenhouse Gas," *Yale Environment Review*, August 4, 2020, https://environment-review.yale.edu/despite-international-efforts-scientists-see-increase-hfc-23-potent-greenhouse-gas.

4. Mario Molina, "The Nobel Prize in Chemistry 1995, Mario J. Molina—Biographical," The Nobel Prize, November 2007, www.nobelprize.org/prizes/chemistry/1995/molina/biographical/.

5. Fiona Harvey, "Mario Molina Obituary," *The Guardian*, October 12, 2020, www.theguardian.com/environment/2020/oct/12/mario-molina-obituary.

6. US Department of State, "The Montreal Protocol on Substances That Deplete the Ozone," https://2009-2017.state.gov/e/oes/eqt/chemical pollution/83007.htm.

7. Lee Thomas, "Clearing the Air About Reagan and Ozone," *Wall Street Journal*, October 9, 2015, www.wsj.com/articles/clearing-the-air-about-reagan-and-ozone-1444338641; Space Center Houston, "Mission Monday: Reagan Launches NASA's Mission to Build a Space Station," Space History Houston, January 20, 2020, https://spacecenter.org/mission-monday-reagan-launches-nasas-mission-to-build-a-space-station/.

8. World Air Quality Index, "World's Air Pollution: Real-Time Air Quality Index," https://waqi.info/#/c/10.047/27.706/2.1z.

9. EXXpedition, "About Us," https://exxpedition.com/about/about-us/.

10. Attribution unknown.

CHAPTER 3: LIVE LIKE CREW, NOT LIKE A PASSENGER

1. Nelson Wyatt, "Cirque Founder Hosts Show for Earth from Space," *Toronto Star*, October 9, 2009, www.thestar.com/news/world/2009/10/09/cirque_founder_hosts_show_for_earth_from_space.html.

2. Aaron Rowe, "Cirque du Soleil Founder Headed to Space," *Wired*, July 4, 2009, www.wired.com/2009/06/cirqueduspace/.

3. One Drop, "Our Impact: Making a Difference, One Drop at a Time," www.onedrop.org/impact.

4. NASA, *International Space Station Benefits for Humanity*, 3rd ed., 89, www.nasa.gov/sites/default/files/atoms/files/benefits-for-humanity_third.pdf.

5. NASA, "About Sustainability Base," www.nasa.gov/ames/facilities/sustainabilitybase/about.

6. NASA Earth Applied Sciences, "Enhancing Water Management," https://appliedsciences.nasa.gov/what-we-do/water-resources; Aries Keck, "When It Comes to Water, You Have to Think Global," NASA, April 20, 2020, www.nasa.gov/feature/when-it-comes-to-water-you-have-to-think-global/.

7. Steve Graham, Claire Parkinson, and Mous Chahine, "The Water Cycle," NASA Earth Observatory, October 1, 2010, https://earthobservatory.nasa.gov/features/Water.

8. NASA Jet Propulsion Laboratory, "Welcome to ECOSTRESS," https://ecostress.jpl.nasa.gov/.

CHAPTER 4: NEVER UNDERESTIMATE THE IMPORTANCE OF BUGS

1. Bob Granath, "NASA Helped Kick-Start Diversity in Employment Opportunities," NASA, July 1, 2016, www.nasa.gov/feature/nasa-helped-kick-start-diversity-in-employment-opportunities.

2. Richard Paul, "How NASA Joined the Civil Rights Revolution," *Smithsonian Air & Space*, March 2014, www.airspacemag.com/history-of-flight/how-nasa-joined-civil-rights-revolution-180949497/.

3. Margot Lee Shetterly, *Hidden Figures: The American Dream and the Untold Story of the Black Women Mathematicians Who Helped Win the Space Race* (New York: HarperCollins, 2016).

4. Society of Women Engineers, "SWE Fast Facts," https://research.swe.org/wp-content/uploads/sites/2/2018/10/18-SWE-Research-Flyer_FINAL.pdf.

5. Elizabeth Howell, "Apollo-Soyuz Test Project: Russians, Americans Meet in Space," Space.com, April 25, 2013, www.space.com/20833-apollo-soyuz.html.

6. Smithsonian, "Bug Info: Numbers of Insects (Species and Individuals)," www.si.edu/spotlight/buginfo/bugnos#:~:text=In%20the%20world%2C%20some%20900,from%20present%20and%20past%20studies.

7. Smithsonian, "Bug Info."

8. Francisco Sánchez-Bayo and Kris A. G. Wyckhuys, "Worldwide Decline of Entomofauna: A Review of Its Drivers," *Biological Conservation* 232 (April 2019): 8–27, www.sciencedirect.com/science/article/pii/S0006320718313636?via%3Dihub#f0005.

9. Damian Carrington, "Fates of Humans, Insects Intertwined, Warn Scientists," *Guardian*, February 20, 2020, www.theguardian.com/environment/2020/feb/20/fates-humans-insects-intertwined-scientists-population-collapse; Pedro Cardoso et al., "Scientists' Warning to Humanity on Insect Extinctions," *Biological Conservation* 242 (February 2020), sciencedirect.com/science/article/pii/S0006320719317823?via%3Dihub#bb0910.

10. Carrington, "Fates of Humans, Insects Intertwined."

11. Alex Ruppenthal, "Fearing the 'Insect Apocalypse'? Renowned Entomologist Says 'Get Rid of Your Lawn,'" *WTTW News*, October 2, 2019, https://news.wttw.com/2019/10/02/fearing-insect-apocalypse-renowned-entomologist-says-get-rid-your-lawn.

12. GrrlScientist, "Scientists Again Warn About Global Insect Decline—but Will We Act?" *Forbes*, February 14, 2020, www.forbes.com/sites/grrlscientist/2020/02/14/scientists-again-warn-about-global-insect-declinebut-will-we-act/?sh=171b3862356a; Dominique Mosbergen, "Insects Are Dying En Masse Risking 'Catastrophic' Collapse of Earth's Ecosystems,"

Huffpost, February 11, 2019, www.huffpost.com/entry/insect-population-decline-extinction_n_5c611921e4b0f9e1b17f097d; Sánchez-Bayo and Wyckhuys, "Worldwide Decline of Entomofauna."

13. Philip Mella, "The Road Not Taken: The Amazing World of Insects," *Pikes Peak Courier*, February 20, 2019, updated June 25, 2020, https://gazette.com/pikespeakcourier/the-road-not-taken-the-amazing-world-of-insects/article_9e1b2c00-2f14-11e9-ab9d-930da7d2cc50.html; Mary Hoff, "As Insect Populations Decline, Scientists Are Trying to Understand Why," *Scientific American*, November 1, 2018, www.scientificamerican.com/article/as-insect-populations-decline-scientists-are-trying-to-understand-why/.

14. Pheronym, Inc., "Space the Final Frontier—for Nematodes," *EINPressWire*, August 10, 2020, www.einpresswire.com/article/523499160/space-the-final-frontier-for-nematodes.

15. E. O. Wilson Biodiversity Foundation, "E.O. Wilson," https://eowilsonfoundation.org/e-o-wilson/.

CHAPTER 5: GO SLOW TO GO FAST

1. Otto Steinmayer, "Desiderius Erasmus: *Adagia* II, 1, 1: Festina Lente," University of Birmingham, Philological Museum, September 12, 2001, http://people.virginia.edu/~jdk3t/FLtrans.htm.

2. Jillian Scudder, "The Sun Won't Die for 5 Billion Years, so Why Do Humans Have Only 1 Billion Years Left on Earth?" PHYS.org, February 13, 2015, https://phys.org/news/2015-02-sun-wont-die-billion-years.html; Eric Betz, "Here's What Happens to the Solar System When the Sun Dies," *Discover*, February 6, 2020, www.discovermagazine.com/the-sciences/heres-what-happens-to-the-solar-system-when-the-sun-dies.

3. "Hertzsprung-Russell Diagram," Cosmos: The Swinburne Astronomy Online Encyclopedia of Astronomy, https://astronomy.swin.edu.au/cosmos/H/Hertzsprung-Russell+Diagram.

4. Matthew Wilburn King, "How Brain Biases Prevent Climate Action," *BBC Future*, March 7, 2019, www.bbc.com/future/article/20190304-human-evolution-means-we-can-tackle-climate-change.

5. Kate Ravilious, "Thirty Years of the IPCC," *Physics World*, October 8, 2018, https://physicsworld.com/a/thirty-years-of-the-ipcc/.

6. "Congressional Testimony of Dr. James Hansen, June 23, 1988" (transcript), www.sealevel.info/1988_Hansen_Senate_Testimony.html; Oliver Milman, "Ex-Nasa Scientist: 30 Years On, World Is Failing 'Miserably' to Address Climate Change," *The Guardian*, June 19, 2018, www.theguardian.com/environment/2018/jun/19/james-hansen-nasa-scientist-climate-change-warning.

7. Intergovernmental Panel on Climate Change, www.ipcc.ch.

8. Jonathan Lynn and Werani Zabula, "Outcomes of COP21 and the IPCC," *World Meteorological Organization Bulletin* 65, no. 2 (2016), https://public.wmo.int/en/resources/bulletin/outcomes-of-cop21-and-ipcc.

9. Association of Space Explorers, "Call to Earth: A Message from the World's Astronauts to COP21," *Planetary Collective*, December 5, 2015, YouTube, https://youtu.be/NN1eSMXI_6Y.

10. Shyla Raghav, "5 Questions You've Wanted to Ask About the Paris Agreement," *Conversation International News*, updated November 4, 2019, www.conservation.org/blog/5-questions-youve-wanted-to-ask-about-the-paris-agreement?gclid=EAIaIQobChMIq-uM8uvm6gIVIgiICR3prggNEAAYASAAEgJ6j_D_BwE.

11. Climate Reality Project, "Why Is 1.5 Degrees the Danger Line for Global Warming?" March 18, 2019, www.climaterealityproject.org/blog/why-15-degrees-danger-line-global-warming.

12. Ed Struzik, "How Thawing Permafrost Is Beginning to Transform the Arctic," *YaleEnvironment360*, January 21, 2020, https://e360.yale.edu/features/how-melting-permafrost-is-beginning-to-transform-the-arctic.

13. Robert McSweeney, "Explainer: Nine 'Tipping Points' That Could Be Triggered by Climate Change," *Carbon Brief*, February 10, 2020, www.carbon

brief.org/explainer-nine-tipping-points-that-could-be-triggered-by-climate-change.

14. McSweeney, "Explainer: Nine Tipping Points."

15. Melody Schreiber, "The Next Pandemic Could Be Hiding in Arctic Permafrost," *The New Republic*, April 2, 2020, https://newrepublic.com/maz/article/157129/next-pandemic-hiding-arctic-permafrost.

16. Esprit Smith, "NASA's Orbiting Carbon Observatory-3 Gets First Data," PHYS.org, July 15, 2019, https://phys.org/news/2019-07-nasa-orbiting-carbon-observatory-.html.

17. Ravilious, "Thirty Years of the IPCC."

18. Chelsea Gohd, "New Climate Report Is Sobering but Strangely Hopeful," Space.com, September 27, 2019, www.space.com/ipcc-2019-climate-report-sobering-hopeful.html.

19. The Audacious Project, "The Nature Conservancy," https://audaciousproject.org/ideas/2019/the-nature-conservancy.

20. Tim Folger, "The Cuyahoga River Caught Fire 50 Years Ago. It Inspired a Movement," *National Geographic*, June 21, 2019, www.nationalgeographic.com/environment/2019/06/the-cuyahoga-river-caught-fire-it-inspired-a-movement/.

21. Lorraine Bossoineault, "The Cuyahoga River Caught Fire at Least a Dozen Times, but No One Cared until 1969," *Smithsonian*, June 19, 2019, www.smithsonianmag.com/history/cuyahoga-river-caught-fire-least-dozen-times-no-one-cared-until-1969-180972444/.

22. Goldman Sachs, "Environmental Market Opportunities: Center for Environmental Markets," www.goldmansachs.com/citizenship/environmental-stewardship/market-opportunities/center-for-environmental-markets/.

23. Marina Koren, "The First American to Vote from Space," *Atlantic*, November 8, 2016, www.theatlantic.com/science/archive/2016/11/voting-from-space/506960/.

24. Koren, "The First American to Vote from Space."

CHAPTER 6: STAY GROUNDED

1. NASA, Glenn Research Center, "Plasma Contactor," in "Powering the Future," September 20, 2011, www.nasa.gov/centers/glenn/about/fs06grc.html#:~:text=Under%20these%20conditions%2C%20the%20space,Plasma%20Contactor%20Unit%20(PCU).

2. Gaétan Chevalier et al., "Earthing: Health Implications of Reconnecting the Human Body to the Earth's Surface Electrons," *Journal of Environmental and Public Health* (January 2012), www.ncbi.nlm.nih.gov/pmc/articles/PMC3265077/.

3. World Health Organization, "Cancer in Children," September 28, 2018, www.who.int/news-room/fact-sheets/detail/cancer-in-children.

4. American Childhood Cancer Organization, "Childhood Cancer Statistics," www.acco.org/childhood-cancer-statistics/.

5. Collaborative on Health and the Environment, "Environmental Contributors to Childhood Cancers," June 3, 2020, www.healthandenvironment.org/webinars/96518; US Environmental Protection Agency, "Research Grants: NIEHS/EPA Children's Environmental Health Centers Center for Integrative Research on Childhood Leukemia and the Environment (CIRCLE)," Research Grants, www.epa.gov/research-grants/niehsepa-childrens-environmental-health-centers-center-integrative-research.

6. Michael Johnson, "Cancer-Targeted Treatments from Space Station Discoveries," NASA, March 27, 2019, www.nasa.gov/mission_pages/station/research/news/b4h-3rd/hh-cancer-targeted-treatments.

7. Deborah Byrd, "How Far Could You Travel and Still See Earth?" *EarthSky*, September 4, 2018, https://earthsky.org/astronomy-essentials/in-space-how-far-away-can-you-see-earth.

8. Frank White, *The Overview Effect: Space Exploration and Human Evolution* (Reston, VA: American Institute of Aeronautics and Astronautics, 1998); NASA, "The Overview Effect," Gary Jordan interview with Frank White, *Houston We Have a Podcast*, episode 107, edited by Norah

Moran, recorded June 11, 2019, www.nasa.gov/johnson/HWHAP/the-overview-effect.

9. Peter Suedfeld, Katya Legkaia, and Jelena Brcic, "Changes in the Hierarchy of Value Preferences Associated with Flying in Space," *Journal of Personality and Social Psychology* 78, no. 5 (October 2010): 1411–1435, https://pubmed.ncbi.nlm.nih.gov/20663027/.

10. Kirsten Weir, "Mission to Mars," *Monitor on Psychology* 49, no. 6 (June 2018), www.apa.org/monitor/2018/06/mission-mars.

11. Baylor College of Medicine, Center for Space Medicine, "Ar/VR 2019 Workshop: Augmented/Virtual/Extended Reality for Improving Health in Space," www.bcm.edu/academic-centers/space-medicine/translational-research-institute/news/arvr-workshop-2019.

12. Catherine Thorbecke, "Confined in a Small Space Due to COVID-19? Here's Some Tips from an Astronaut," *ABC News*, April 9, 2020, https://abcnews.go.com/Health/confined-small-space-due-covid-19-tips-astronaut/story?id=70018745.

13. Matt Johnston, "Australian Antarctic Division Launches Virtual Space Station," *iTnews*, October 10, 2018, www.itnews.com.au/news/australian-antarctic-division-launches-virtual-space-station-513761.

14. Lewis Gordon, "Can Virtual Nature Be a Good Substitute for the Great Outdoors? The Science Says Yes," *Washington Post*, April 28, 2020, www.washingtonpost.com/video-games/2020/04/28/can-virtual-nature-be-good-substitute-great-outdoors-science-says-yes/.

15. James Kingsland, "'Virtual Reality Nature Boosts Positive Mood," *Medical News Today*, October 21, 2020, www.medicalnewstoday.com/articles/virtual-reality-nature-boosts-positive-mood.

16. David Delgado and Dan Goods, "The Studio at NASA's Jet Propulsion Lab," SCI-Arc Channel, December 6, 2018, https://youtu.be/8rGZGuacQ9s; NASA, Jet Propulsion Laboratory, "The Studio at JPL," www.jpl.nasa.gov/thestudio/.

CHAPTER 7: WHATEVER YOU DO, MAKE LIFE BETTER

1. American Geophysical Union, "COVID-19 Lockdowns Significantly Impacting Global Air Quality," *Science Daily*, May 11, 2020, www.sciencedaily.com/releases/2020/05/200511124444.htm#:~:text=They%20found%20that%20nitrogen%20dioxide,the%20same%20time%20in%202019.

2. Lucy Handley, "'We Can't Run a Business in a Dead Planet': CEOs Plan to Prioritize Green Issues Post-Coronavirus," *CNBC*, August 10, 2020, updated August 11, 2020, www.cnbc.com/amp/2020/08/10/after-coronavirus-some-ceos-plan-to-prioritize-sustainability.html?__source=instagram%7Cmain.

3. World Economic Forum, Global Future Council on Space Technologies 2019–2020, "Six Ways Space Technologies Benefit Life on Earth," briefing paper, September 2020, www3.weforum.org/docs/WEF_GFC_Six_ways_space_technologies_2020.pdf.

4. GoodReads, "Larry Niven: Quotes: Quotable Quote," www.goodreads.com/quotes/16687-the-dinosaurs-became-extinct-because-they-didn-t-have-a-space.

5. Caltech, "Space Solar Power Project," www.spacesolar.caltech.edu/.

6. Debra Werner, "Aerospace Corp. Calls for Collaboration in Space Solar Power," *Space News*, October 27, 2020, https://spacenews.com/aerospace-space-solar-power-collaboration/.

7. Timothy "Seph" Allen, "Sentinel-6 Michael Freilich to Stand Vigilant of Rising Sea Dangers," NASA, November 20, 2020, https://appliedsciences.nasa.gov/our-impact/news/sentinel-6-michael-freilich-stand-vigilant-rising-sea-dangers#:~:text=%22Earth%20Science%20shows%20perhaps%20more,for%20NASA's%20Science%20Mission%20Directorate.

8. Michael Sheetz, "The Space Economy Has Grown to Over $420 Billion and Is 'Weathering' the Current Crisis, Report Says," *CNBC*, updated October 2, 2020, www.cnbc.com/2020/07/30/space-economy-worth-over-420-billion-weathering-covid-crisis-report.html?ct=t(04102020_TheBridge_AnsariAnniversary).

9. World Meteorological Organization, "Essential Climate Variables," https://public.wmo.int/en/programmes/global-climate-observing-system/essential-climate-variables.

10. XPrize Foundation, www.xprize.org.

11. Charlotte Trueman, "Why Data Centres Are the New Frontier in the Fight Against Climate Change," *Computer World*, August 9, 2019, www.computerworld.com/article/3431148/why-data-centres-are-the-new-frontier-in-the-fight-against-climate-change.html.

12. George Carlin, "The Planet Is Fine," Genius, https://genius.com/George-carlin-the-planet-is-fine-annotated.

CONCLUSION

1. Mote Marine Laboratory and Aquarium, "Mote's Restored Coral Spawns on Florida's Coral Reef," August 12, 2020, www.youtube.com/watch?v=Xsgphgb0-JU&feature=youtu.be.

2. Charles Adams, Leslie Sturmer, and Alan Hodges, "Tracking the Economic Benefits Generated by the Hard Clam Aquaculture Industry in Florida," FE961, University of Florida, IFAS Extension, October 2014, https://edis.ifas.ufl.edu/pdffiles/FE/FE96100.pdf.

3. Florida Keys National Marine Sanctuary, "Coral Reefs Support Jobs, Tourism, and Fisheries," https://floridakeys.noaa.gov/corals/economy.html#:~:text=By%20one%20estimate%2C%20coral%20reefs,full%20and%20part%2Dtime%20jobs.

4. NOAA Office for Coastal Management, "Fast Facts: Coral Reefs," December 15, 2020, https://coast.noaa.gov/states/fast-facts/coral-reefs.html.

COURTESY OF THE AUTHOR

NICOLE STOTT is an astronaut, aquanaut, mom, and artist who spent over one hundred days in space aboard the International Space Station and on the final flight of the Space Shuttle *Discovery*; her NASA career spanned nearly thirty years. Nicole painted the first watercolor in space, which is now featured at the Smithsonian. She founded the Space for Art Foundation to unite a planetary community of children through the awe and wonder of space exploration and the healing power of art. She is on a mission to inspire everyone's appreciation of our role as crewmates here on Spaceship Earth. Nicole shares this message with audiences around the world, including at the Vatican and the United Nations' historic Paris Agreement gathering, and she has been featured in *National Geographic*'s documentary series *One Strange Rock*. An instrument-rated pilot and avid diver, Nicole lives in Florida with her husband, son, and two dogs.